D0252946

Instability Rules

Instability Rules

The Ten Most Amazing Ideas of Modern Science

Charles Flowers

John Wiley & Sons, Inc.

Published by John Wiley & Sons, Inc., New York
Published simultaneously in Canada

ISBN 0-471-38042-3

Printed in the United States of America

10 9 8 7 6 5 4 3 2 1

For A. R. Casavant, Lucille Johnson, and Bill Alfred,
teachers who tried to make us use our minds.

Contents

Preface

"It Moves . . . "

Eppur si muove . . . Did a bitter old man, disgusted with the scarlet-robed, fig-faced dolts around him but desperate to hang on to life, mumble that phrase under his yellow breath, after kneeling and recanting what he had discovered through his spyglass? That's the story, said not to be true, but it ought to be.

Pah! Earth doesn't move around the Sun, the Inquisition forced the Pisan astronomer to say in Holy Rome. No, Earth is the center of the universe, put there by God and fixed until the end of time. Orthodoxy: right thinking.

And yet, that grumbling *Eppur si muove* . . . something like, "Dammit, it does too move!"

For generations of students, Galileo's apocryphal but irresistible muttering affirms a heliocentric solar system: Earth and the five planets known in the early seventeenth century circling around the Sun. Suddenly, sensible humans grasped what had not been understood during the 100,000 years or more we have, according to current analyses of the genetic record, been here on Earth in our present form: The senses through which we apprehend existence are frail and treacherous. Bright Phaeton in his swift chariot, Quetzalcoatl careening over ancient Tenochtitlán, are still. It is we who tumble around them, unequipped to sense this truth, feeling the ground firm beneath us.

But the old man's wisecrack knocks ajar another door of perception that did not fully swing open until the startling discoveries of the twentieth century: that there is perpetual movement everywhere and in everything where we sense stability, not just in the journey of the Earth.

Rapid, continual change . . . imprecision . . . an infinite number of perspectives: Nothing is fixed.

For the next 300 years after Galileo, science was essentially the continuing, expanding discovery that the universe is ever larger, more complex, and more ancient than the experience of our senses would suggest. Yet the Sun stood still.

Then we learned, as this book will recall, that the solar system is moving at preposterous speed, as is our home galaxy the Milky Way, and likewise the supercluster within which it is a grain of sand, and, indeed, everything in the universe—racing away from an inconceivable explosion that began time some 12 billion to 15 billion years ago.

We learned that the continents are forever slipping and sliding around the globe, like clothing on a teenager, and the mountains are forever rising, the oceans widening, the volcanoes stoking their furnaces for the next blast.

Our bodies are a fever of change as our minds perpetually rewire themselves and our genes make uncountable decisions, renewing or growing or misfiring to produce the runaway cancers that may kill us, initiating the instability of mortal decay.

We learned that entire galaxies are being born billions of light-years away in deep space, or being ingested into the supposed oblivions of black holes, or blowing themselves to smithereens.

Within tiny atomic universes, particles pop in and out of being, impossible as that may be to conceive, while atoms collide and meld, buzzing continually in their electrically charged states.

This, then, was the truth behind many of the defining discoveries of the twentieth century: existence is constant activity.

It is a counterintuitive theme . . . and many of us hate it. Who are we if we are not the same physically and mentally from nanosecond to nanosecond, unable even to freeze-frame our face in the mirror or solidify for a moment our feelings about a lover. It's all moving too fast. The present is instantly the past. What sermons can be found in nature, since it is in perpetual flux? What the hell, then, is "truth"?

The twentieth century was perhaps predicted by the Greek thinker Heraclitus's ancient dictum, often quoted by conceptual artists in the 1970s, "You can't dip your hand into the same river twice." But the idea goes much further than that now. It's also not the same hand, the same owner of the hand, the same riverbank, the same Sun in the sky, the same universe.

I write in late August in a converted barn in Westhampton, New York. Under a Giotto-blue sky, single florets of the amethyst phlox at the edge of the woods occasionally fall to the ground in a gentle breeze. The Japanese maple that blunts the north wind has begun its submission to autumn. Aside from my irregular arpeggios on the laptop, and the far-off whine of what I take to be some jerk's cigarette boat, there is only a growing wash of sound, like the muted roar of a seashell, that is the wind rising before a late afternoon rain.

All is still. I am in the center of my world. All is illusion. There is no center anywhere of anything.

1

Hubble and the Expanding Universe

There may be, as we learned in school, no straight lines in sprawling, liquid nature, but there are rigidities and mechanisms that have the force of rules. Clocks tick all over the place. Apparent movements in the sky inspire our hours and calendars. Radioactive rocks decay, clicking off billions of years of terrestrial geology. Pulsars, superdense neutron stars, spin hundreds of times a second, sending out radio beams as regularly as a downed airliner's black box. (The record so far is 860 rotations per second.) Light travels at a cosmic speed limit of 186,282.397 miles per second (a speed so phenomenal that liberal arts–trained copy editors, I've found, routinely change it to "per hour"); the distance it travels in a year becomes the space-time measurement, a light-year. If we look at an object only 1 light-year away, the light that reaches us has traveled 5.9 trillion miles and, perhaps more astonishingly, is conveying information that is twelve months back in the past. This is highly accurate distance-timekeeping.

Over time these regularities will become less so, perhaps even the unvarying speed of light, but they work more than well enough for humankind's fleeting appearance in the universe.

One of the most valuable regularities for astronomers is a star called the Cepheid variable, a yellow supergiant that is typically a thousand to a hundred thousand times brighter than our Sun. Each brightens and dims in a predictable cycle ranging from 3 to 50 days. This cycle is related to the star's brightness. In other words, the brightness of a Cepheid, linked with the brevity or length of its cycle, reveals its distance from the Earth. Once thousands of Cepheids had been cataloged, the absolute brightness of a star with a 10-day cycle, say, could be estimated. Comparing this brightness with its apparent brightness in the night sky would produce an estimate of its distance. Cepheids can be used as signposts in space—or what scientists call "astronomical yardsticks" or "standard candles."

These extremely helpful celestial objects became the specialty early in the 1900s of Henrietta Swan Leavitt, an astronomer who pored over photographic plates counting and cataloging thousands of individual stars at the Harvard College Observatory in Cambridge, Massachusetts. Her work produced signposts that were very carefully calibrated. They were like loran way points for the yachtsman navigating a dark sea, or VORs for the amateur pilot, fixed in relation to the Earth, defining points of space. (As of this writing, thanks to observations by the orbiting Hubble Space Telescope, Cepheids have been cataloged up to 56 million light-years away.)

One astronomer who understood Cepheids, and just about everything else in the night sky that was deepening because of advances in optical telescopy in the 1920s, was the inimitable, insufferable Edwin Hubble. An American midwesterner who affected a British accent and claimed to have a German dueling scar—the kinds of eccentricities that are more amusing to read about than sit beside—he was an

Edwin Hubble

athlete, though no team player, whose ambitions aimed, well, to the stars.

When Hubble lit up his briarwood pipe and peered into the 100-inch-wide Hooker Telescope on Mount Wilson in southern California in the 1920s, the most sophisticated astronomers knew oceans less than you do about the night sky. They knew that some stars must be hundreds of millions of light-years away, and they knew there were hundreds of thousands of them. These were heady advances in knowledge made possible in the eighteenth and nineteenth centuries by the growing efficiency of telescopy and related devices. They were so heady, in fact, that, as you might expect, a snarky grad student can find scores of remarks by great names in the field that can sound pleasingly fatuous to someone facing orals: "We have learned everything there is to learn about the cosmos" is the theme.

Every age is arrogant in its own way. To be fair, they barely had time to note down, much less understand, their new knowledge. Consider that for a hundred thousand years, whether our earliest forebears knew how to count them or not, a human being could see only eight thousand to nine thousand stars on a moonless night. These stars, even down to Galileo's era, seemed to be affixed permanently—with the exception of the Sun, Moon, and five visible planets—on the inside of a great sphere that rotated around our fixed homes, wherever they might be, on Earth. That sphere had to be pretty large, for the hot Sun and glowing Moon revolved within it, but at the same time it felt comfortingly close.

Binocular vision. You use it automatically; your guinea hen or iguana cannot. When you close either eye, your field of vision covers a different area from when you close the other eye. With one eye closed, you lose your ability to guess near distances accurately; you aren't seeing in 3-D. The stars are so far away, of course, that human binocular vision cannot possibly gather the stereoscopic information necessary to site them.

But as it orbits the Sun in a year, the Earth moves some 186 million miles from one extreme to the other every 6 months. The common sense of two centuries ago suggested that observations from the same spot at 6-month intervals would show a star appearing to move in relation to its neighbors. This displacement of a nearby star against the backdrop of stars in deeper space is called stellar parallax.

But the predicted phenomenon was not immediately observed. It became clear that the stars were so extravagantly far off that even an imaginary giant creature with eyes set 186 million miles apart, a macrocosmic Godzilla, would not have the binocular vision required to guess the distance of the nearest star, whatever that might be. It was not until 1838 that a star, 61 Cygni, was observed closely enough to be sited; Friedrich Bessel reckoned it to be at the then astonishing distance of 61 trillion miles. (For the record, the nearest star is Proxima Centauri, an in-our-face 24.8 trillion miles away.)

Despite this calculation and many that followed, the professional watchers of the skies still operated under one overwhelming assumption in the 1920s: Our galaxy was the entire universe. The vast legions of stars and other celestial objects all lay within the Milky Way.

Hubble, like many people who expand the history of science, had a peculiar infatuation, professionally speaking. While mainstream astronomers concentrated on the apparent Everests—giant stars, comets, planets, novae, and supernovae—he fixated upon something as evanescent and humble as oil slicks on a mud puddle. Clearly visible to the unaided eye are faintly glowing patches of light, Wite-Out smudges, powder trails on the ebony bowl of night. Until the mid-nineteenth century, speculation clustered around two possible explanations for these glowing clouds, called nebulae (the ancient Latin plural for nebula, meaning vapor or mist). Perhaps they were just what they looked like, fogs of cosmic gasses and sparkle dust. Or perhaps, now that better telescopes and cameras were continually discovering more far-off stars to count, they were exceedingly distant groups and gatherings of stars.

The answer to the debate was disappointingly anticlimactic: both. It was found by using a new device, the spectrograph, that would marry chemistry to astronomy and provide a wealth of information about the prosaic nature of celestial objects. Much as a sheer curtain of tiny rain droplets separates the afternoon sunlight into shimmering, arced bands of color, the spectrograph breaks down the radiance of a star into thin lines of light and dark color once it is funneled through a telescope to the device. Each celestial object has a characteristic spectrum, or range of these assorted lines, each spectrum as unique as a living creature's DNA.

For those devoted to the deeper mysteries of the universe, the spectrograms provided a sweeping revelation of the universal mundane: no celestial wonder, no matter how dramatically mammoth or intriguingly bizarre, was found to have an element that does not exist right here on homely Earth. The same building materials are used in every perceptible object. Even today, when such previously unimaginable phenomena as black holes and quasars and the rest are subject to spectrography, no exotic *Star Trek* element has been found.

And the spectrograms nailed the nebulae. A third of them turned out to be stars in throngs; the other two-thirds were mists of dust and gas.

These nebulae were the indistinct smudges that held Hubble's attention during long nights in the cold mountain air at Mount Wilson in the 1920s. Did he look where few others looked long because he knew he was singular, his absorption as characteristic as the faked posh accent? Did he sense that there are infinite riches in the humblest phenomena? Regardless, he won the lottery and defined the limits of our galaxy for the first time in human experience.

In the constellation Andromeda is a spectacular nebula, aptly known as the Great Nebula. First recorded in 905 and named "little cloud" by Al-Sufi, an ancient Persian astronomer, it was found in 1912 to have a spectrum more like starlight than gases. By taking long exposures of photographic plates with the Hooker Telescope, Hubble found that this nebula is a galaxy separate from the Milky Way that corrals hundreds of billions of stars in its huge spiral shape, which is 150,000 light-years in diameter. Gazing at the Great Nebula on a clear night, you are seeing the only heavenly phenomenon outside our home galaxy that can be seen with the naked eye.

And within it Hubble found twelve telltale Cepheid variables, their brightness linked with periodicity showing it to be 800,000 light-years off in space. Actually, we now know that Andromeda, also called M31 by professional astronomers, is 2 million light-years from us, but Hubble's achievement was not the measurement but the concept.

A fifth of the way into the twentieth century, he found, as no one had before him, that our galaxy is only a part of the universe, not the universe itself. He found that stellar distances were much greater than had been imagined by most.

But the faraway stars, the newly met neighboring galaxy, seemed just to hang there, fixed in space.

The cosmos was about to rock. Heraclitus's ever-flowing river, the ancient metaphor, took on vast new meaning as Hubble looked at his spectrograms and found that nothing is fixed. Stars, solar system, galaxies—even the Cepheid way points are not stable coordinates, not even for a fraction of a second.

Why this frantic hustle? In what direction or directions? And, of course, as we always ask when pondering in amazement the latest discovery about the nature of our illusory universe, to what end? It would be another 30 years before the first two questions would be answered—perhaps decisively—and the last is still debated.

But astronomy had enough to chew on when Hubble proved that the entire universe is engaged in a mad dash. The evidence was blatantly obvious in each celestial object's spectrum.

Spectrography not only analyzes chemical composition, it also records changes in light or sound waves that reveal cosmic speed.

To understand how this speed detector works, we have to think briefly about the electromagnetic spectrum, which ranges from high-energy gamma rays at one extreme to low-energy radio waves at the other. This spectrum covers all of the electromagnetic radiation that courses through the entire universe. In succession from highest energy to lowest, the different types are gamma rays, X rays, ultraviolet radiation, visible light, infrared radiation, microwaves, and finally radio waves. Each of these types of electromagnetism is defined by the length of its waves and the frequency of their vibrations—ranging from the short, high-frequency gamma rays through the other types of waves to the much longer, much slower radio waves, with wavelengths up to six miles long. Visible light, which takes up only 2 percent of the electromagnetic spectrum, is near the center.

As a light-emitting celestial object moves away from us, its light waves appear to lengthen, which moves them toward the slower end of the electromagnetic spectrum, where the light waves are longer, and causes them to turn red. This "redshift" can be used to determine speed as well as the direction of the object's movement in relation to us. If the object is racing toward us, its light waves will appear to become shorter and will turn blue, or have a "blueshift."

This effect is usually likened to a similar and perhaps better-known phenomenon of sound waves, the Doppler effect. Sound waves will seem to shorten as music, blasting from a passing car, say, approaches, and therefore the sound rises in pitch, then the waves lengthen as the performance speeds off in the distance, falling proportionally lower in pitch as the car moves away.

Hubble knew that his Andromeda nebula was getting closer to our home galaxy, thanks to a blueshift observation made in 1912 by American astronomer Vesto Melvin Slipher. Of course, from Hubble's work with the Cepheids he now knew that an entire galaxy of swirling stars, not just a luminous gas cloud, was trucking closer. (Actually, we and Andromeda may be nearing at about 50 miles per second.) Meanwhile, Slipher had been studying other supposed nebulae (all later to be identified as galaxies), finding that most were redshifted—that is, moving away from us—and some were flying off at speeds as fast as 700 miles per second.

Slipher's observations were significant, but it was Hubble who used analyses of redshifts and blueshifts to make his second astonishing observational and conceptual breakthrough. He and his colleague Milton Humason recognized that there is an invariable relationship between the redshift of a galaxy and its distance from us: The greater the shift (in other words, the greater its velocity), the farther away it is.

The implications were quite literally mind-boggling, although schoolchildren and addicts of the Discovery Channel take them for granted today. First, all galaxies were on the move, including some that could be measured to be zooming off at a seventh the speed of light. Second, the most distant were moving the fastest. Five years after discovering that the universe has many galaxies, Hubble discovered that it was also, as he announced in 1929, filled with cosmic movement.

And the essential characteristic of that movement, as the redshift showed, was that all matter is racing away from all other matter: The universe is expanding.

When Hubble recognized that tiny spectrum shifts spoke of vast stellar congregations moving at incredible velocities, the relationship he found between a galaxy's distance and speed became famous as Hubble's Law.

In 1956, some three years after his more celebrated colleague's death, Humason and collaborators revised Hubble's Law to incorporate new information suggested by the theory that the universe began 12 billion to 15 billion years ago in a mysterious explosion known as the big bang. Possibly, galaxies sped apart even more swiftly in the early years of their formation but have slowed since.

But are they flying apart now at a constant rate? If so, there should be a "Hubble constant," and calculations backward in time should be able to pinpoint when the big bang occurred—in other words, tell us how old the universe must be. Scientists do not yet agree on the value of this constant.

For that reason, estimates of the age of this particular universe (yes, there may well be others raging in and out of existence) vary so widely, perhaps irking and confusing nonscientists. In daily life, an estimate of 12 billion to 15 billion, or 10 billion to 20 billion in the conflicting estimates of some theorists, would not be very practical. In cosmic terms, however, the disparity of a few billion comes close to being a distinction without a difference. Also, an agreed-upon range chucks out large areas of possibility at each extreme.

Recent attempts to ascertain the value of the Hubble constant, appropriately relying upon observations of the most distant known Cepheids by the Hubble Space Telescope, may suggest that the universe is somewhat more youthful than most experts have assumed. Measurements seem to indicate that other galaxies are hustling away from us at 180,000 miles per hour for every 3.26 million light-years of their current observable distance. If the light reaching us from a galaxy suggests that it is twice that far, or 6.52 million light-years, then it will be moving out at 360,000 miles per hour.

If this estimate is fairly accurate, why can't we just count backward? Because the density of mass in the universe would affect the rate of movement of celestial objects as it expands, and that density has not been determined. Also, because several extraneous factors might distort the recent observations, the estimate has to be considered indecisive.

So here we go again. If the Hubble constant is as low as 180,000 miles per hour, the universe may be 8 billion to 12 billion years old, but if that constant is as great as almost three times higher, as at least one esteemed astrophysicist and Hubble acolyte contends, then the age rises to 20 billion years or so.

The kinds of things Hubble learned from light—that slender 2 percent band of the electromagnetic spectrum—are now accessed from sources of information up and down the entire spectrum.

For example, many objects that do not emit light do radiate X rays, from black holes to quasars, pulsars to stars and galaxies. Most X rays, which have shorter wavelengths than light and vibrate more frequently, do not make it through the Earth's protective atmosphere and can be captured only by space-based X-ray detectors. *Uhuru,* the first such satellite, was sent up by NASA in 1970. In its more than two years of data transmission it sent back word of more than three hundred previously unknown celestial objects.

Similarly, other orbiters and space voyagers are gleaning information from lightless sources that emit gamma rays, ultraviolet waves, infrared waves, and radio waves. It is not from visible light but from cosmic gamma rays that huge sources of antimatter have been found. As the term suggests, antimatter is the opposite of matter. Very few particles of antimatter, which theoretically should exist in the same numbers as the particles of matter that form almost the entire detectable universe, have been found on or near the Earth. But in 1997, antimatter was found spurting up in a plume 35,000 light-years tall and 4,000 light-years wide at the center of the Milky Way, emitting a steady stream of gamma rays detectable only by an orbiting satellite-observatory. Gamma rays are produced either in nuclear reactions or from the decay of radioactive elements. The discovery of this gargantuan fountain might be evidence of a black hole, perhaps the turbulence in space-time as it sucks in celestial objects that wander too near, or the gaseous residue of stars in their explosive death throes.

So at the beginning of the twenty-first century, scarcely 75 years since Hubble's first transformative discovery, your tax dollars are supporting probes into space—that is, you are contributing to the accelerating accumulation of knowledge about the nature of the universe—with detectors of stellar messages in forms that we could never discern visually, no matter how sophisticated the optical telescope.

But that device, too, is still being used and to even better effect, upgraded in efficiency as a result of the computer revolution. The light captured optically during the long, cold nights at Mount Palomar, say, is now converted by silicon chips, or "charge-coupled devices" (CCDs), into highly detailed digital images that are transferred onto magnetic tapes and then viewed and analyzed at a computer terminal by an astronomer coming to work at regular business hours and presumably drinking the day's second cup of coffee as he sits back in his ergonomically correct chair waiting for the computer to boot up. No danger of freezing eyelashes stuck to the eyepiece.

It is impossible not to believe that these platoons of sophisticated devices roaming the entire electromagnetic spectrum will not soon make discoveries that will astonish us even more than our great-grandparents were flabbergasted to learn that our galaxy was not the universe, only an undistinguished suburb, and that all of existence had no stable resting point, no harbor still and quiet.

Already, in the tradition of Hubble's work, the continuing discoveries ratify not only the insignificance of our solar system, galaxy, and galaxy group but also the prevalence of rampaging change and instability everywhere known.

Within the apparent boundaries of an expanding universe we continue to discover discrete if huge cataclysmic events. The Hubble orbiter has captured a galaxy in the act of devouring another galaxy. (Yes, Andromeda *does* have a blueshift.) Stars much larger than our Sun end their days dramatically in huge explosions, supernovae, first documented by the Chinese in 1054. Recent observations of one such cosmic blast show that it was as wide as ten thousand solar systems. The Hubble found evidence in 1996 that the universe has at least fifty billion galaxies, five times the estimate at the time. At the same time, great clouds have been recognized as the nurseries of new stars and galaxies. Constant change, constant movement, constant life and death.

We will return to the vexing question of whether the universe will expand infinitely, finally losing its energy and becoming a vast, cold, dead sphere of burned-out or asphyxiated celestial objects, or take some other tack.

But our solar system and galaxy may not be around that long. Rather than end in fire or ice, we might simply get conked on the head, all of us at once, and never be the wiser. No warning, no pain. Gamma-ray bursts, the most intense sources of energy yet discovered in the universe, explode all around the space-time continuum. From as far away as halfway across the universe, 8 billion light-years away, they explode with more power than the Sun has produced in its entire working lifetime. Some three hundred of these huge blasts are detected annually, lasting from a few thousandths of a second to a few minutes. Forget about Hollywood's vagrant asteroids. The unpredictable, implacable flash of a gamma-ray burst, blasting toward us at the speed of light, would wipe our galaxy clean before we sensed a thing.

Why not? We are seeing stars explode to smithereens in innumerable legions, galaxies colliding with each other to form quasars, and vast curtains woven from billions of stars hanging across space billions of light-years away. The common discovery of the dozens of orbiting observatories, including the Hubble, is that the serenity of the night sky is turmoil incarnate, an ongoing tale of violent births and violent deaths. In comparison, the clockwork movement so difficult for some to accept four centuries ago—the peaceful scudding of the Earth and other planets around our Sun—has the comfort of home.

2

Einstein and the Wonder of Light

Gamma-ray bursts, or lesser jolts, such as the comet or asteroid that destroyed the dinosaurs about 65 million years ago, were not what Albert Einstein meant by saying "God does not play dice." He wanted to believe that it all makes sense, even as he explained how a fourth dimension profoundly complicates the question of where we are.

If it seems that Mozart fell out of his mother's womb whistling four-movement symphonies, it is just as likely that newborn Albert Einstein, first getting his world into focus, looked at the light—such as it was in Ulm, Germany, on March 14, 1879—and felt the prevocal equivalent of *"Was ist das?"* Unlike Mozart's, however, his genius was hidden and unacknowledged for decades, though obviously simmering from the beginning.

Light obsessed him. As he would say later, he kept asking children's questions long past puberty, the kind that drive parents bananas. "It's blue . . . because, it's blue! Get over it, Al. It's just . . . light." He explained once that he asked children's questions because he was "a late developer. . . . I first pondered such 'simple' questions as an adult, and so probed them more deeply and tenaciously than any child would do."

Albert Einstein

Evidently distant from family and friends, but generally amiable enough, the young Einstein was perhaps the only person on the planet who incessantly thought about light. The rest of humankind pretty much took it for granted, as far as we know. Even other geniuses of physics, such as Isaac Newton, seemed able to discuss the properties of light—how it breaks down into a rainbow spectrum of colors when passed through a prism, how it seemed to be a stream of tiny particles beamed at an observer—and go on to other things. With Einstein, the need to understand light was a chronic condition.

For about the past 80 years or so it has been fashionable among some academics to discount the role of the individual in momentous discoveries. To be sure, if there had not been the unique product of nature and nurture known as Einstein, someone else, or even an academic committee, eventually would have copped to his theories about relativity, Brownian motion, and all the rest.

If there had never been the Mozart we know, there might probably have appeared another musician—perhaps Salieri!—who would bridge the gap in the classical period between late Haydn and early Beethoven. But it is impossible to believe that this ersatz Mozart would have written the heartbreaking recognition scene at the end of *The Marriage of Figaro,* in which we all forgive and are forgiven, or the irresistible "Twinkle, Twinkle, Little Star," which with apparent simplicity as brilliant as Occam's razor survives all vulgarizations and musical gentrifications. By somewhat the same token, our understanding of relativity and its implications, of space-time, would be much poorer without the wit, tenacity, humility, conviction, and showboating of Einstein. To recognize such individuals as unique is not to deny to the prodigious forces of history and thought their heft. Let us take comfort in our singularities when we can.

The bedrock of Einstein's work is twofold: the special theory of relativity, which he announced in 1905, and the general theory of relativity, published exactly a decade later. Perhaps, before looking at these revolutionary discoveries, we can scrape away some of the unhelpful encrustations they and their discoverer have accumulated.

In 1919 Einstein became overnight what he is now—the symbol, occasionally tongue-in-cheek, for colossal intelligence. He had predicted that light traveling through space would bend when passing a massive object with a strong gravitational field. An expedition of English scientists led by astronomer Arthur Eddington, measuring starlight during a total solar eclipse, found that certain stars seen near the darkened Sun seemed to change position. In other words, the Sun's gravitational field bent the light as it passed near, making the stars appear to move, from a terrestrial observer's point of view.

Quite a bit of nonsense ensued in the popular imagination. Einstein was portrayed as the quintessential absent-minded professor. (He added more than a brushstroke to this portrayal himself, I believe. He had to know that that tumbleweed hair set him apart in Princeton in the 1950s, and a man who makes the following statement in the United States is not unaware of his audience: "The hardest thing in the world to understand is the income tax.") His theories were considered to be the product of such inconceivable intelligence that only seven other people on the planet could understand them. His insight that time is the fourth dimension was taken to have mystical implications. Not many ordinary human beings thought it worth their while to try to understand what the man discovered.

But consider this story, often told by Harvard science historian I. B. Cohen: A publisher decided that someone, somewhere, would have both the mathematical competence and the storytelling ability to write a book about Einstein's theories for the general audience. A competition was announced. The manuscripts came rolling in. The authors' names were attached in sealed envelopes, so that the judging would not be influenced by academic connections and the like. Presumably, as in any pile of manuscripts, a great number were written by lunatics and many by professors who lacked the common ear. But one slim manuscript, however, sparkled among the others like a jewel of brevity, wit, and clarity. The envelope, please. . . . Yes, it was written by Albert

Einstein, illustrated by his own drawings. The "smartest man in the world" could talk to the people.

"Everything must be made as simple as possible but not one bit simpler," he once said. In reading about relativity, it may help to imagine our mind, like our brain, divided in halves. With one, we can glimpse the beauty and the mystery of certain fundamental concepts; with the other, we can take the complex equations, the inexpressible ambiguities, the unanswered questions, and file them for our next reincarnation.

The special theory of relativity is based upon the realization that time is indeed that fourth dimension, after height, length, and width. This is a practical—not a mystical—concept. The three familiar dimensions define the physical world we see. They locate us, or so it seems.

But time is just as much an indicator of position if we consider measurements on a scale larger than Earth. When and where are intertwined, because of the nature of light.

An example: Light from the Sun takes about 8 minutes to reach the Earth, and vice versa. If the Sun suddenly clicked off at 10 in the morning in your time zone, you wouldn't know until 10:08. And if some kind of observer were able to perch on the Sun's cinder, waiting to see what happened down here, the sight of our panicked actions wouldn't reach the dead star until 10:16, your time.

When did the Sun go belly up? It depends upon where the observer sits, and time is the fourth dimension that pinpoints the event. In this case we would know about an event 8 minutes after it occurred, but from the point of view of the observer on the Sun, we seemed to learn about it 16 minutes after it occurred. When did the Sun expire? When were we affected? Depends upon where you were.

Until Einstein, it was reasonable to assume that time was a fixed yardstick. We could count from the present to the past along a calibrated line. When the great Danish astronomer Tycho Brahe saw a supernova flare up in 1572, that was the date of the event. Now we know that such celestial events may have occurred millions or billions of years ago, but the light of the explosion has just reached us. We have to know how far away it is to know when, by our clocks, it occurred. Time is the fourth dimension measuring the event as a physical occurrence—nothing New Wave about it.

We could no longer think of time as a fixed quantity. If, as many experts believe, there might well be a solar system like ours relatively near—say, only 100,000 light-years away—the sentient creatures there might have devices capable of detecting the first small band of *Homo sapiens* roaming in wonder over the East African savanna. The aliens in their present see our distant past.

But by the time a videocassette of the birth of humankind made it back for airing on the Learning Channel, at speeds perhaps approaching but never able to reach the speed of light, the humans of Earth or their descendants would have evolved, if still in existence, and the date would be some years past A.D. 102,100. In other words, each celestial object, star, or possible solar system has its own time. We can know the time only in our own backyard.

Actually, such an exchange may have been initiated in 1974, when a simple image was beamed from a radio telescope in Puerto Rico toward the hundreds of thousands of stars in the cluster known to astronomers as M13. Any alien civilizations on planets orbiting any of those stars "now" have 21,000 years to develop the technology to capture and interpret this message. If they replied instantly, we would not receive their reaction until at least forty-one Times Square millennium blowouts from now—42,000 years to get the answer to "Is anyone out there?" And an affirmative reply would be 21,000 years out of date.

But measurements of time and space are further complicated by the expansion of the universe, the apparent distaste of most visible matter for most other visible matter—an insight that inspired Einstein to conceive of what we now know as "the space-time continuum."

It's important to remember that Einstein, in contrast with most scientists familiar in the popular imagination, did not reach his most important conclusions by performing lab experiments. To the amazement or the annoyance of others, he seemed to grasp a startling concept intuitively, know it to be true, then come up with imagined examples to prove his point. He didn't like to explain this thought process in detail, and perhaps he couldn't.

In any event, his thought experiments have become fairly familiar. Many were addressed to this question: Since it was generally accepted by the end of the nineteenth century that light was a kind of wave, what would it be like to travel along this fleet ripple?

Beginning in 1895, when he was 16 years old, Einstein began worrying incessantly over the problem of traveling 186,400 miles per second (the accepted speed of light in his day) holding a mirror. This was an adventurous conceptual problem, but he believed—precisely the right verb here—that it could be solved only with common sense. Again, this warning may be necessary: Bizarre as his discoveries might seem, they are insights into the physical world, not messages from the supernatural or the metaphysical.

In one experiment, the teenager imagined a traveler with a mirror on a train barreling along at the speed of light.

Choosing to assume that the speed of light is constant, because that just made sense to him (and apparently not even considering whether anyone else would disagree or have to be convinced), Einstein pondered what the mirror would then show. If the passenger faces the caboose, his reflection would have to travel faster than the speed of light to bounce back to his eyes. Why? Because he is racing forward at the speed of light, and his reflected image is generated at a distance to the rear of his position. But Einstein was convinced that it is impossible for anything to travel faster than the speed of light—that is, for the reflection to catch up to the passenger. That being true, would the mirror then be blank? No; that defied common sense, his chosen arbiter.

But what about someone observing this fantastic train from a railway platform? He would be able to measure—remember, this is a thought experiment—the speed of the train as equal to the speed of light. But if the traveler was regarding his visage in the mirror, the light from the mirror would appear, to the outside observer, to travel at twice the speed of light. That couldn't be.

Mulling this and related problems for a decade, Einstein came to the counterintuitive conclusions that produced his special theory of relativity. Others, most notably French theoretical physicist Jules-Henri Poincaré, sensed that the time was ripe for some kind of paradigm shift in theory. Three years before announcing his discovery, Einstein had read this startling contention in the Frenchman's work: "There is no absolute space. . . . There is no absolute time." But it seems clear that no one else was bearing in on the problem with Einstein's conviction and creativity.

His conceptual leap makes perfect sense in words but does not immediately "compute" in terms of everyday experience. For one thing, the effects he describes do not occur until the train or other objects attain velocities near the speed of light. For another, common sense leads to the revelation that reality is, well, surreal.

Begin with this simple formula, which can be explained very briskly by any highway patrolman: Speed equals distance divided by time. If the radar detector catches you driving 90 miles per hour, your rate of speed would, if maintained constantly, carry your vehicle 90 miles within the next hour.

Don't skip lightly over this apparently self-evident notion. It works in the everyday world, including traffic court, because it is based upon the observable fact that the distance and the time in the equation are absolute. No single hour is shorter or lengthier than any other hour, according to our agreed-upon system of time, and every 90-mile-long stretch of highway is precisely the same length as every other 90-mile-long stretch of highway.

We know that. But we have to set it aside, and, like Einstein, see what happens when we look at that simple formula—speed equals distance divided by time—if distance and time are *not* fixed.

Do not despair. At this point Einstein himself said, "I used to go away for weeks in a state of confusion, as one who at that time had yet to overcome the state of stupefaction in his first encounter with such questions."

The idea of relativity begins simply: If you are moving smoothly at a steady speed, you don't know how fast you're moving, or even that you *are* moving except in relation to something outside. In TV ads, the manufacturers of expensive cars would have you believe that a dozing passenger will experience no sense of movement because the ride is so quiet and so smooth that the outside world does not intrude. Similar principle. Not many people outside the Northeast ride trains these days, but if you are the exception, you've undoubtedly had the experience of not knowing whether your quiet diesel is beginning to move or the train across the platform is moving. If your car windows are closed and the radio is blaring, you might rarely have the same odd sensation at a stoplight.

Let's return to that passenger left to stare glumly for a decade of young Einstein's life at the pesky mirror. According to this basic concept of the relativity of perceived motion, the mirror has to beam back his reflection. Otherwise the little world inside the train—passenger, mirror—would have evidence, *without* looking outside the window, that it was traveling at the speed of light. With the shades drawn, looking at his image in the mirror, traveling on an imaginarily noiseless train, the passenger could imagine himself stationary just by looking in the mirror.

This explanation, which satisfied Einstein and led to greater leaps in reasoning, will satisfy you only if you accept the basic principle, to state it again, that you can't know how fast you're going without making reference to something else.

But what about the other participant in Einstein's thought experiment, that observer at the station? To him, wouldn't the light from the mirror appear to be moving at twice the speed of light—the sum of its speed from mirror to passenger plus the speed of the train past the station?

No. Einstein relied again upon common sense and upon assumptions that seemed logical.

Speed has to be distance divided by time.

If the speed of light is the same to the passenger and that layabout in the station, despite the obvious differences in their physical circumstances, then something had to be true that had probably never been considered seriously in physics: distance and time, each of them considered fixed in the normal course of events, must both somehow be susceptible to change; otherwise the formula doesn't work. Approaching the speed of light, both of the elements considered stable in our world of experience are dramatically altered. Once again, Einstein assumed that a basic principle—speed equals distance divided by time—was writ in stone even as new information suggested that the principle had surprising, counterintuitive implications. The equation must hold true; to save it, new views of the nature of time and distance are required as we approach the speed of light.

In describing his special theory of relativity in 1905, Einstein explained two fundamental principles: (a) The speed of light remains the same whether it is beamed from a stationary or a moving light source, and (b) the light will appear to move at that speed to either a fixed observer, like the man on the station platform, or a moving observer, like the passenger in the train.

Whether you are lounging in a beach chair or rocketing up into orbit in a space shuttle, the speed of light will be the same to you. It is the universal speed limit, unchanging. It applies not only to light, that 2 percent of the electromagnetic spectrum, but also to all waves along the entire electromagnetic spectrum, from the fevered gamma waves to the phlegmatic radio waves.

The light from a supernova, the X-ray evidence of a ravenous black hole, the "live" coverage of a guerrilla attack halfway around the world—none of these events reaches us at the instant they happen. There is the delay required for the electromagnetic waves to travel from event to observer. What may appear instantaneous is in fact an event that, to be fully understood, has to be understood in terms of the dimension of time.

Accepting these ideas leads to further puzzling thought experiments.

Let's abandon those bygone trains for a bit of *Star Wars.* Suppose we have invaded an enemy galaxy in our Stealth rocket ship, shielded like an astronomical blackbody so that no light radiates or is reflected off us. Nonetheless, our wily adversary somehow knows that we are near. Their commander, seated at a battle station in the exact center of a disk-shaped starship, orders that laser beams be fired in all directions, just on the chance of hitting us.

From his perspective, the entire fusillade will occur simultaneously. His orders will reach each cannon at the same time, and the light from their blasts will ricochet to him simultaneously.

But if the starship is moving past us as it fires, we will see something quite different: the laser at the point farthest from us will seem to fire later than the laser at the point nearest us. The speed of light is constant; it takes longer to travel a greater distance. From our perspective, then, the lasers were not fired simultaneously.

What we have experienced, if our shield has survived the assaults, is the relativity of simultaneity. In other words, the firings were simultaneous to the captain; we, observing from a different *frame of reference,* did not see them as simultaneous.

Which frame of reference is accurate? It's relative.

So, too, when frames of reference are moving in relation to each other, distance and length are relative. As our adversaries sailed by, an enemy aide walked parallel to us carrying a computer analysis of the results of the firing. From printer to alien commander's chair was a walk of, say,

10 feet to him, but it looked longer to us because the craft was moving past. Did he *actually* walk 10 feet, or more than 10 feet? Again, it depends on the frame of reference.

But surely observers in both frames of reference can agree on the length of the ship as it passes by. No. To measure this apparent length, we have to choose a time, then mark where both the fore and the aft of the ship appear to be in relation to our own ship at that time. It is quite possible, Einstein realized, that the vast ship could be 100 feet in length and yet, when we measure the distance between the points marked on our own ship, the length is different.

It is the nature of thought experiments that certain aspects of reality are discarded in order to concentrate on the main point. In this case we must now imagine that we can hear a clock ticking on the enemy craft. To the commander, the clock works normally. To us, because the clock is moving past us as we hang poised in space, it seems to tick more slowly.

Such imaginary observations are virtually impossible to replicate in our everyday experience, but with Einstein's equations it is possible to calculate what strange things happen when one frame of reference moves at velocities close to the speed of light.

This is the basis of the numerous documentaries that depict rocket ships of the future traveling away from Earth and back at almost the speed of light, only to find that time has moved much more slowly on the moving ship than on the (relatively speaking) stationary Earth. The astronauts emerge some years older, but their friends are aged or long dead. In fact, this *time dilation,* as it is called, can be estimated. At 99 percent of the speed of light, the time on the speeding craft would become a seventh of its value on Earth; the astronaut who experienced a 10-year trip would return home to a planet and its inhabitants that would be 70 years older. Also at that speed, the ship would weigh seven times more than when it was stationary on the ground. Time slows down, weight increases, and length grows longer as an object approaches the cosmic speed limit.

For these reasons, the speed of light is a speed limit. Einstein imagined an example of an electron, one of the subatomic particles that whirl around the nucleus of an atom, accelerated to or past the speed of light. As the electron is somehow pushed so that it accelerates near the velocity of light, the clocks in its frame of reference are slowing down. That means that the force behind the acceleration necessarily acts upon the electron for increasingly shorter periods of time. At the speed of light, as another equation devised by Einstein proved, the acceleration of the electron in its moving frame of reference becomes zero. It moves at the speed of light, but no faster. Extending this line of reasoning, he realized also the increasing difficulty of accelerating the electron—that is, the increasing amount of energy expended upon pushing it—meant that it was getting heavier.

This recognition led him to conclude in 1905 that "the mass of a body is a measure of its energy content." Or, to translate the concept into the equation that virtually everyone knows, $E = mc^2$.

That is to say, *E,* the total energy in a specific amount of matter, is the same as *m,* its mass, multiplied by *c* squared—the speed of light squared.

The renowned equation shows that mass and energy, though existing in different forms, are essentially the same thing. In order to give off energy, an object loses its mass, in the proportion described in Einstein's formula. The concept is true of a burning lump of coal, a love handle, or an exploding supernova.

It is easy to accept the implications of this equation. It is a bit harder to grasp firmly Einstein's various observations about phenomena near the speed of light, unless we accept his assumptions, which are a mixture of common sense and intuition and fairly complex mathematical equations that renovate the equations of earlier physics. But perhaps the most important challenge for the general reader is to grasp the definition of the "space-time continuum" that arises from Einstein's 1905 theory.

Accepting time as the fourth dimension of the physical universe, we don't know where something is until we know when it is, and vice versa. No event happens simultaneously with any other event; all are separated from each other by the speed limit imposed upon the electromagnetic spectrum. The enemy starship commander and we, the lurking intruders, were both right about the firing of the lasers, although our different frames of reference caused us to come to different conclusions about time and distance.

Space-time can be graphed, sort of, on two-dimensional paper, but that may be the wrong approach to conceptualizing it, for most of us. Keeping the idea in words rather than insufficient pictures works best for me. Since it is difficult to visualize a fourth dimension in the three-dimensional world we think we inhabit, why try? But we can understand the concept by considering how it works.

Visualization works very well for earlier theories. That's why we all think we understand gravity—it's the thing that caused the apple to fall on Isaac Newton's head. (Yes, this incident is actually a fable, passed along after his death by

Newton's spinmeister of a niece to French philosopher Voltaire, but it works fine in the classroom and in our memories.) In fact, no one understands gravity, despite that clear-cut image. Einstein spent the last decades of his life trying to work it out. He failed. It's still one of the abiding mysteries of science.

Whatever it may be and however it works, gravity was redefined by Einstein ten years after his paper on the special theory of relativity in the second of his astonishing landmark works, the general theory of relativity.

This is the theory that was affirmed by those stargazing Brits in 1919. The Sun seemed to bend the light beamed from distant stars. In fact, as Einstein had predicted, the Sun's mass curves space-time itself. The light from the star followed that curve.

This observation, brilliantly simple, completely transfigures centuries of theorizing about the mechanics of the universe, beginning with our tiny, homely solar system. In ancient Greece, transparent spheres with planets and stars attached to them were thought to circle around the Earth, and then the Earth and other planets were proved to circle around the Sun, thanks to Galileo, and then the galaxies were clocked at huge velocities moving away from each other, thanks to Hubble. But all of these movements were thought to occur because forces, even if unknown, must be acting upon objects, causing their revolutions or mad dashes.

Not so, according to Einstein's new thought experiments and equations. Gravity did not hold the Earth in its orbit because it was a force reaching out from the Sun to keep us on a leash. Instead, gravity was that phenomenon by which the massive Sun can create curves in the fabric of space-time. Less massive objects, such as Earth, roll around such a curve just as skateboarders in competition careen up and down curved walls.

You will see illustrations of this idea that raise more questions than they answer, as I've noted, but people learn in different ways and perhaps these odd-looking maps of curved space-time will be helpful to you.

Einstein, typically, had reasoned his way to this redefinition of gravity because of dissatisfaction with the scope of his special theory. The 1905 paper was limited because it dealt only with frames of reference—train or starship, steamboat or elevator—that moved at a constant velocity in relation to another frame of reference. But what about situations in which the frame of reference is accelerating rather than moving at a constant speed?

Eager to return to the safety and comfort of the Milky Way, our Stealth spacecraft begins accelerating. We can't see out; because of relativity, we know how fast we're going only because of the information provided by our instrumentation. Suddenly a bright ruby laser beam pierces the wall above our head but curves down to exit the opposite wall at waist level. (This is a thought experiment; we aren't hit and we have very sharp eyesight.)

By chance, the enemy has blasted us, but the surprising phenomenon we just saw was not a consequence of alien technology. The light should have traveled in a straight line, no matter where it entered or exited, but its path was curved. In the terrestrial world, according to the laws of Newtonian gravity, light has to travel the shortest distance between two points—that is, in a straight line.

Einstein, after a realization that he later called "the happiest thought of my life," concluded that acceleration, which would bend the light as our spacecraft gained speed, is the same thing as gravity, which would predictably bend the starlight passing near the Sun.

This insight, this "happiest thought," required years of intricate mathematical calculations before Einstein was ready to publish the general theory of relativity. Because of their complexity, and because the entire notion of a curved space-time invisibly warping the heavens was so counterintuitive and revolutionary, even physicists could not immediately recognize what Einstein had achieved.

Most of us imagine we would prefer a week of root canal work to attempting to follow the logic of even the simplest of the equations that affirm Einstein's theories, but we can familiarize ourselves with his conclusions. And we can marvel, I think, at two running themes of the story: He always trusted the logic of the math, no matter how wildly the implications contradicted the conventional wisdom formed over millennia by the greatest minds of science; and somewhat to the contrary, he continued to rely upon assumptions that made sense to him, even if they could not be proved.

The truth is, most people with an interest in the story of relativity and even a byte of memory of high school algebra, not to mention a skill at fantasy football, can walk through some of Einstein's formative equations. The experience may take time, but it will confirm the central truth: The logic of math is the proof of the seemingly illogical principles of relativity.

Those who recall photos of Einstein sawing away at his violin ("perfectly correct, totally uninteresting," reported one accompanist) may forgive me for recalling the powerful theme in the last movement of Beethoven's sixteenth string quartet, sometimes rendered as the question *"Muss es sein?"* and supposedly answered *"Es muss sein!"* (Must it be? It must be!)

Only once did Einstein betray that clarity of conviction, and it led him to publish an insight he later regarded as the blunder of his life. Remember, his two great theories were proposed before Hubble discovered that the universe was expanding. For once Einstein was as much a prisoner of conventional wisdom as anyone else, sharing the assumption that the universe is static. According to his work with gravity, however, the universe had to be dynamic—that is, either contracting or expanding.

To make his equations work, he theorized the existence of some unknown force, which he called the "cosmological constant," a kind of antigravity needed to keep the universe from increasing or expanding in size. Hubble's discovery did away with the need for this quick fix.

Fade in to the present and a different problem: The rate of expansion calculated according to relativity doesn't make sense in terms of the amount of mass known to exist in the universe and the time required to create the galaxies we see. No one is questioning relativity, but the cosmological constant has been dusted off and brought back to the table. Einstein's insight, though a possible solution to a different problem, turns out not to be a blunder—at least not by the definitions most of us use. It is difficult, once again, to separate the man from his music or, as Yeats put it, "the dancer from the dance."

If Hubble's work reveals that the galaxies are flying apart, Einstein's reveals the instability of fundamental dimensions in our perceptually stable world. We can only know what time it is close to home. When we see a celestial object, we are looking backward into time. Great invisible forces are shaping and reshaping the universe, whipping space-time like a blanket.

To touch another human being, to see another human being, requires the conduction of electromagnetic waves, and so, however infinitesimally small in time it is, there is always a gap between us. One has to wonder if that has something to do with the essential loneliness of the human condition.

Be that as it may, Einstein and most other physicists of his generation soon accustomed themselves to the peculiarities of relativity. The equations made sense. Relativity produced certainties, even if they were counterintuitive.

But he would never accept, as the majority of the community of science has, the existential instabilities proposed by Niels Bohr concerning the foundation of everything, the atom. Einstein believed in rules; once we travel inside the atom, we can no longer make predictions about the behavior of the very matter that forms us. We can only come up with probabilities.

As you read, subatomic particles in your body are madly defying common sense. Describing how they behave put a tremendous strain on the classic approach of theoretical physics, as expressed by Nobel laureate Robert Laughlin: "Physics teaches us that rules dreamt up without the benefit of physical insight are nearly always wrong. Correct rules must be discovered, not invented." Even today, the line between discovery and invention may not always be indisputably clear in quantum physics, and that is entirely appropriate to the subject.

3

Bohr and the Puzzles of the Quantum World

Niels Bohr, according to the familiar stories, does not come across as the standard-issue, puckish Dane. Athletic, reserved, and steady, a brilliant theorist with a learning disorder whose marriage meant that his wife could take over the task of writing out his papers from his mother, Bohr is generally portrayed as a man who discovered the strangest secrets of existence and wrestled soberly with their implications. He is also a man who taught himself English by reading the novels of Charles Dickens, which might not seem to be the shortest or the easiest path to the goal.

Grave or not, he may have found all the humor he needed in the world he illuminated: the subatomic world of quantum physics, where virtually no event or entity obeys the physical laws of the classical Newtonian world in which we move, pitch softballs that do not defy gravity, pretty much remain fat and 50 day after day, and get caught out by our significant others when we try to be in two places at once.

Quantum physics, quantum mechanics, subatomic world—these terms are at first daunting, and the concepts they describe are incomprehensible to everyone. Repeat: no one understands their fuzzy realities, and no one pretends to. Yet unlike relativity, which makes sense but is often hard to hold on to conceptually, quantum physics is actually quite

Niels Bohr

simple to describe, even as the actions it encompasses are puckish, strange, and even frightening.

"Subatomic" essentially covers the parts within the atom: the electrons, the protons and the neutrons found in an atom's nucleus, and numerous smaller particles. "Quantum mechanics," so named for a reason that will become obvious, refers to their laws of behavior.

The theory originated in the work of physicist Max Planck, who announced in a 1900 lecture that light and other forms of radiant energy were not continuous waves but streams of tiny, discrete amounts, which he called "quanta." Moreover, he found that this concept could explain the mysteries of blackbody radiation, a pesky problem for classical physicists. All radiant energy beamed upon a blackbody, an ideal substance, would theoretically be absorbed and none be reflected. (The closest approximation in the real world, lampblack, reflects less than 2 percent of incoming radiation.) Bewilderingly, classical physics could not predict the frequencies of radiation that would subsequently be emitted by the theoretical blackbody. Planck devised a predictive formula, but only by fixing the energy frequencies to certain values, instead of the expected continuum of values. He couldn't explain why this would be so, but it worked in his theoretical analysis of the problem. This was only the beginning of the dislocation that most professionals and certainly the wider public would feel with the discoveries of quantum theorists. Light was not a continuous beam but a stream of birdshot; in fact, Einstein would call these quanta "photons" some 5 years after Planck's lecture. Heat radiating from a fireplace on a chill day was now a series of discrete levels of energy, not the comforting flow of warmth we experienced.

No greater test of common sense than quantum physics has yet arisen. Planck himself was suspicious of his discovery. No branch of knowledge has so far proved so resistant to the signature human question "Why?" It doesn't make sense, but it explains, or at least accurately describes, events that cannot be explained or described by any other means.

One of the most frequently abused scientific phrases in casual discourse is "quantum leap," popularly employed to mean something like Chairman Mao's "great leap forward." It is important to see how the quantum leap, Bohr's major discovery, is something quite different from the common usage.

For this concept, we change direction. Rather than rocketing off in space at speeds near the speed of light, we drill into the atom. Less than a century ago, in 1911, Ernest Rutherford announced his astonishing discovery that every solid object is made almost entirely of empty space. This means you. The current estimate of the number of cells in the human body is a cool thirty trillion. Each of those cells is constructed from about ninety trillion atoms. Each of those atoms that is you is almost entirely composed of empty space.

Rutherford's experiments showed that each atom has a tiny nucleus, or center, with even tinier electrons (with a weight one/two-thousandth as much as a proton or neutron) orbiting around it at distances that are vast by subatomic standards. For example, if an illustrator were to draw an atom with a nucleus ¼ inch in diameter, so that we could see it, the orbit of the electron whizzing around farthest from the nucleus would be more than 2,000 feet across. As someone has suggested, this is like a pea in the center of a circle as wide across as seven football fields. (In other words, you've never seen an accurate rendering of an atom.) The nucleus, for all of its mass, takes up only one billionth of the atom's total space.

Rutherford also discovered that the nucleus has a positive electrical charge, evidently balanced by the combined negative charges of the atom's electrons, and contains almost all of the atom's total mass.

He was exactly right on all counts, as we know now, but he was a shy young researcher from New Zealand working in England, the mother country, and the Old Bulls of science at Cambridge University dismissed his theory out of hand. Proud heirs to Isaac Newton, they knew that Rutherford's atom would smash itself flat, according to the great man's laws of physics. The revolving electrons, radiating energy as they spun, would grow weaker and be pulled to the center by the positive, stable, comparatively massive nucleus.

Indeed, each electron would fall out of its orbit, according to Rutherford's model, within a fraction of a second. And if that would necessarily occur in any one atom, it would occur in all atoms. The entire universe would collapse into a landfill of tiny atomic nuclei.

Two years later, Niels Bohr dealt Newtonian physics its second body blow, following Einstein's 1905 theory of special relativity. And there was more than a coincidental relationship. In Einstein's first thought experiments, he accepted the traditional concept of light as a wave. By 1905, however, he had agreed with Max Planck that light is actually broken up, as are all waves on the electromagnetic spectrum, into little bits. (In fact, his paper explaining light quanta was published at the same time as his paper on special relativity.) Each of these tiny, tiny, tiny pieces of light was called a quantum, meaning simply that it is a discrete, separate quantity of something. In the case of light, the quantum is known as a photon. These tiny particles do not change, cannot be changed. They are basic.

Only two years after Rutherford's announcement, Bohr suggested that the electrons of an atom do not find themselves pulled in a curve toward the nucleus because they cannot move toward it in a continuous line. Just as the quantum of light is a kind of discontinuity in nature, so the electron's existence is governed by a discontinuity: it can exist only in certain places, certain clearly differentiated energy states, which he called "stationary states," and each of those is a clearly differentiated orbit.

Once again, of course, common sense is offended as much as the theories of classical physics. Bohr himself said that "the theory is crazy." Why can't an electron shift in a continuous line from one orbit to another? The Moon's orbit, for example, could move somewhat farther from the Earth (indeed, it creeps away from us at the rate of about 1 inch a year), the orbit of Earth somewhat closer to the Sun, but the electron can only make a quantum jump, a quantum leap, seeming to disappear from one orbit and instantly appear in another. How can this be? *Es muss sein.* How did it get from one place to the next? It just did. And this is a very different phenomenon, as suggested earlier, from the cocktail party conception of quantum leap. This type of leap is prodigious not because of its breadth but because of its unaccountable and even unacceptable nature.

Furthermore, as the electron leaps from one stationary state to another, it will emit or absorb radiation. Based upon the electron structure of the simplest atom, hydrogen, Bohr could predict what its line spectra would look like.

In other words, the man's peculiar theory brought resolution to a problem that was vexing to physicists. Back in the nineteenth century, remember, spectrographs were used to identify elements in celestial objects. But knowing that hydrogen, say, has a certain spectrographic signature raises an obvious question: Why?

Quantum theory answered that question, in just about its only concession to the human desire for explanations. Bohr came to recognize that the quantum leaps of electrons change the bar codes of a spectrograph: When an electron pops into a lower orbit, a bright line appears in the atom's spectrographic fingerprint, and a dark line is registered by the spectroscope when an electron is shifted into a higher orbit. Since any element is uniquely identified by the number and spacing of the electrons around its nucleus, the spectrograph's sensitivity to the quantum leap signatures made for a reliable Identikit. (For this reason, although the unaided eye cannot confirm it by looking up at the night sky, the most common color produced by celestial objects is a medium shade of red. When an electron orbiting a hydrogen atom, the most common element in the universe, makes a quantum leap to a higher state and back again, it emits a red-colored photon.)

We seem to be back on firm ground here. Bohr cannot explain why electrons are assigned to a fixed number of possible orbits, but his insights do explain why an objective scientific instrument would obtain consistent results.

Still, the Old Bulls were even less amused with Bohr than with Rutherford. If they had long ago put childlike questions behind them, it was the nature of adult questions in science to hone in on the whys and the wherefores, to ascertain and enumerate the laws of a phenomenon—in other words, to find the logic in nature.

In addition, there were several mathematical problems that seemed to question Bohr's model on its own terms, but a veritable swarm of creative physicists attacked these complex issues, step by step affirming his fundamental intuition.

It turned out that the model would not work for all atoms unless the researcher could calculate four separate items: the size of the electron orbit; the shape of this orbit (an ellipse rather than a circle); the direction of the orbit; and the electron's spin, which could be clockwise or counterclockwise. These four characteristics, each assigned a quantum number or value, defined a single electron's energy level. According to something called the "exclusivity principle," an electron with a specific set of four quantum numbers occupied a specific orbit, or stationary state. Other electrons that came along in the same state were forced to move up to the next higher energy state.

Add more work by many other theorists, and quite a few equations, and the atom we get is the so-called Quantum

Mechanical Model—not a soccer ball with BBs buzzing around it at great distances, but a kind of pulsating globe of energy, still mostly empty of matter but frenetic with inexplicable activity. This was an astonishing compilation of theoretical and experimental advances, especially when you consider that the proton had only been discovered in 1919, and the neutron in 1932.

It was also astonishing, in retrospect, that so many physicists leaped headlong and headstrong into the quantum soup, eager to tackle the mathematical mysteries afloat there. Bohr led and is known to encyclopedists as "the father of quantum theory," but the others, all of them profoundly creative individuals, guessed that he was correct in the theory, if not in its misty particulars and particulates. It was a heady time for this select few—the womanizing Schrodinger, the mountaineering Heisenberg, the soccer-playing Bohr (a usually competitive jock who occasionally fell into reveries and forgot to watch for the ball).

But their discoveries in this quantum business were not immediately accepted in all quarters. "This is just a cheap excuse for not knowing what's going on," harrumphed one of Newton's heirs. Bohr was careful to protest, "Every sentence I utter must be understood not as an affirmation but as a question." By contrast Einstein would later describe Bohr's creative speculations as "the highest form of musicality in the sphere of thought."

Remember that quantum theory requires that all of existence be divided, without explanation, into two worlds—the macrocosm above the level of the atom, and the subatomic microcosm. Quantum mechanics, with its apparent chaos, rules from within the atom down to the tiniest bits of nature. Classical physics and relativity—both rulebound—may describe or confine the universe, all the way from atom to farthest galaxy or quasar or black hole. It is difficult for many to accept that the quantum world should not work like all of the rest of existence.

On the other hand, a friend's 9-year-old son told me that Bohr's initial insight about "stationary states" is really very easy to understand. "It's like an express elevator," he patiently explained, "that stops only on certain floors." (Yes, he has one of those teachers who can turbocharge lives.)

Another generation Xer of Bohr's day, 23-year-old Werner Heisenberg, came up with mathematical proof in 1925 that Bohr's concept of subatomic particles performing quantum leaps between fixed orbits was demonstrably accurate. Few physicists could follow the logic of Heisenberg's difficult algebraic equations, but after Einstein's similarly challenging math, this particular snag was not necessarily a deal breaker. The problem for some, however, was that the math worked its logical magic but did not make possible the construction of a visual model of the atom. The equations didn't make a picture.

Then yet another young theorist, Erwin Schrodinger, tapped directly into the dementia of the subatomic world: If electrons were treated mathematically as waves, not as particles, his equations also explained the structure of the atom. But how could both explanations—Heisenberg's particles and Schrodinger's waves—be correct?

In science, as in the household, two explanations for a single event are much worse than one. In fact, the dissonance can be devastating.

It was impossible to conceive that electrons could, according to the two different sets of equations, act either like particles or like waves. Surely they had to be one or the other. Particles, more or less like hard little BBs, bounce off each other and, however slight their mass, will exert force in such interactions. This is one class of physical behavior, according to traditional physics. Waves, on the other hand, like the sun-dappled ripples in a swimming pool, can pass through each other, or neutralize each other, or combine to create larger waves. This is another class of behavior, again according to traditional physics.

All of this theorizing occurred quite rapidly. It was only 1926 when Max Born devised the solution that is still used today: There is no solution.

Well, not quite. But Born argued that it would be impossible to find the exact answer about the electron's state. Instead, we can determine only the probability, the likelihood, of that state. The wave used by Schrodinger in his equations does not exist physically, like the waves of light or radio in the electromagnetic spectrum. It is a concept: the range of possible positions for the electron particle.

But Heisenberg would delve into stranger matters still. He made sense of the wave/particle conundrum by showing, in effect, that sense was not the issue.

The quantum of light, you will recall, is called a photon. When we see an object in the world above the boundary of the atom, photons bombard that object and are then reflected back to our eyes. When the Sun conked out in chapter 2, the dying of the light was the termination of the solar photon stream toward Earth.

It can't work that way in the subatomic world. Let us imagine a single infinitesimal photon bombarding a single infinitesimal electron (that's the nature of the interaction at that level). The photon will bounce back to provide visual information about the electron, but at the same time, it will bump the electron out of its position. We can picture the BBs here. When the photon returns, it is an unreliable messenger. It announces where the electron was when they encountered, but not how far away it bounced. A second photon might retrieve that information but also cause a change in the electron's position. In fact, researchers discovered that the wavelength of visible light is too long to site an electron. Gamma rays, electromagnetic radiation with a shorter wavelength, are used, but a shorter wavelength means greater energy: Gamma-ray photons knock the electron even farther away from its original position.

Where, then, is the electron? We don't know. We can't know. We know where it was, and we can make guesses about its possible alternate locations, based upon mass and velocity in the interactions. Heisenberg devised an equation that predicts the degree of uncertainty in any simultaneous measurement of a particle's position and momentum. His work has gone into pop culture as the Uncertainty Principle, and in that milieu the essential idea is often abused.

Heisenberg's was a tedious uncertainty if you wanted to believe that the physical world has incontrovertible laws and that accurate measurements and analyses of actions at every level of physical experience are always possible. And then Heisenberg realized in about 1927 that this new puzzle was an answer—of a very unorthodox sort—to the previous puzzle about electrons. If you couldn't know exactly where the electron particle happened to be, you could plot its possible positions as a wave. In other words, so far as mathematical equations in the human world are concerned, the electron in the subatomic world is either or both a particle and a wave. And it is not that detection is inadequate; no, it's that "reality" is uncertain.

How can it be either or both? Because it acts as if it's either or both. You may not like that answer, although we've had about three-quarters of a century to get used to it, and Einstein hated it. Remember, relativity may be difficult to grasp, but it does have rules and predictable outcomes. Heisenberg's conclusion implies that the activities of electrons and other subatomic particles are not predictable. You can predict a range of outcomes but not a single outcome.

Bohr made some sense of the issue by deciding that an electron behaves like a wave if you examine it with a wave detector, but like a particle if you use a particle detector. And you have to examine it with both types of device to know the electron's properties. His notion, known as complementarity, means that two mutually exclusive behaviors must be studied and somehow accepted as coexistent. Furthermore, an atomic system is not defined until it is examined; until then, there is only a range of probabilities and potential values.

At this point we have to recall one of science's most famous interchanges. If you ever suspected that religion is irrelevant to the mind and heart of the scientist, the following dialogue provides at least a filament of evidence to the contrary.

> EINSTEIN: Quantum mechanics is very worthy of regard, but an inner voice tells me that this is not the true Jacob. The theory yields a lot, but it hardly brings us any closer to the secret than the old one. In any case, I am convinced that He doesn't throw dice.
>
> BOHR: It's not our business to prescribe to God how He should run the world.

Of course, allusions to religion confuse rather than solve a scientific problem, as the story of Galileo showed, but scientists, walking out of the lab into the imbroglios and bewilderments of real life, can yearn for meaning like anyone else.

That was, and is, an underlying concern for many with the existential acrobatics of the quantum world: Without the potential for solid answers, is it possible that there is fundamentally no "meaning"?

Einstein again: "Science without religion is lame, religion without science is blind." This, of course, does not mean he would agree with Red Sox outfielder and theologian Carl Everett, who has announced that dinosaurs never existed because they are not mentioned in the Bible.

"A poem must not mean but be," wrote bourgeois poet Archibald MacLeish. Whatever your religious or metaphysical convictions, it's not unusual to expect the universe to "mean," not just "be." The meanings, the whys, are a bear in quantum mechanics.

Electrons can be in two places at once. Protons can pass through barriers that are physically impermeable even in the subatomic world. Subatomic particles can theoretically communicate with each other across millions of light-years of space at speeds greater than the speed of light. What they say is presumably of interest only to them—turn to the right, reverse course, change spin—but if the message gets through so quickly, is some kind of physical transmission defying the cosmic speed limit? Not only can the electron be either a wave or a particle; it also can "choose" its state of being. And it may not exist at all unless someone or something observes it. The electron may remain somehow inexistent until a human, an electronic device, or even a sheet of mica takes note. Really far-our physicists speculate that the universe we know exists as it does only because it is observed by humans; all of its other possible states of existence—and they might be infinite in number—collapse when we record it.

It's not only that we don't begin to know answers to these marvels but that we know that answers may not exist. Here's the difference: no one knows, although there are a couple of theories, why the full Moon is ten times brighter than a half Moon. This seems odd. But scientists believe that there must be a physical explanation. Possibly, although I always forget to ask, there is a simple reason why they call the wind "Mariah." These mysteries seem to fall into a class of conundrums that can be solved. The mysteries of quantum physics, by contrast, often hover in a cloud of ambiguity. In every observation, there will be a quantum of uncertainty.

And that means, to certain hearts and minds, that the universe itself is irrational, built upon uncertainties and unexamined potential states rather than upon solid matter rigorously following rules of behavior. It is not just that Heisenberg recognized that we cannot observe with certainty the position and momentum values of a particle; it is that the two values are indeterminate whether we are trying to measure them or not. This is not a reasonable state of affairs.

And if the universe is irrational in its tiniest building blocks of matter, it seems likely to many that there cannot possibly be a cosmic plan, divinely inspired or just spinning along. If there is no plan, no actions are predetermined. Chance plays a pivotal role in collapsing waves of possibility in the operations of molecules and galaxies, proteins and neurons.

The implications for physics and metaphysics, as previously noted, are huge, even if debatable, for consequences of actions are not predictable but only probable within a range of possibilities. The effect on everyday speech and thought, while not so fundamental, is nonetheless toxic. The uncertainty principle does not apply to human behaviors unless we're trying to create alibis or making lame jokes. Its implications do not bleed out into our Newtonian lives. While there's no reason to get too exercised about goofy popularizations of a scientific concept, it's worth repeating that uncertainty, as discovered by Heisenberg and refined or debated since, is a theory describing small-scale physical phenomena, specifically one of the house rules of the quantum world.

In one famous lab experiment with photons, familiar to physics students as the double-slit experiment, light is beamed at a screen through two parallel slits. Like the water agitated on the surface of the pool, light should behave like a wave: coming through the two separate slits, the waves would splash against the screen, forming a pattern of dark and light bands, called an "interference" pattern, that results from the action of the waves as they pass through each other, flatten each other out, or combine to make a stronger wave. The dark sections in the pattern are known as destructive interference, light sections as constructive interference. In fact, the experiment yields such patterns. It makes sense.

But a sister experiment also yields such patterns, and results in seeming nonsense. If a single particle of light, a photon—that is, a particle, not a wave—is beamed toward the slits, it would presumably slip through one or the other and hit the screen at a single point. It doesn't. In fact, photons aimed one at a time appear on the screen over time in an interference pattern, as if they were waves, not particles. Why would they do that? Moreover, a photon can "choose" to go through both of the slits—in other words, "be" in two places at the same time.

These contradictions cannot be resolved, it seems, by making a distinction between waves and particles. Light will be defined, if ever, as something that is neither one and is not both.

You don't have to know much about publishing to know that cat books race off the shelves; the creature is the most popular household pet in the United States. By coincidence, in the popularizing of quantum physics, the number one thought experiment, the subject of many of its own books, is Schrodinger's cat. Schrodinger thought it up as something of a scherzo, but it continues to fascinate scientist and nonscientist alike. (And it may have a certain ignoble appeal to those who prefer dogs.)

Schrodinger, far from being pleased that his insights became incorporated in quantum theory, found it absurd that others came to believe that reality in the quantum world could be determined only by probability.

In his thought experiment a live cat is placed inside a sealed box that can become a lethal gas chamber. A radioactive source will decay; as it does so, the decay in its radioactivity is registered by a Geiger counter. When a decay particle is produced, the counter trips a hammer that breaks open a glass flask containing the poison fumes. In the world above the quantum, in the world of Newton's physical laws where apples fall upon heads, the cat suffocates to death.

Hoping to prove the absurdity of quantum probability theory, Schrodinger argued a case in which the theory suggests that there is a 50 percent probability that the radioactive source will decay by one particle per hour. In other words, at the end of that hour there is a 50 percent chance that the cat is dead, a 50 percent chance that it is still alive. Echoing the theory that two wave functions are superimposed, according to quantum probability, Schrodinger noted that that cat would be both alive and dead, until someone observed it.

Since this made no sense, particularly to the cat, Schrodinger believed that he had exposed the absurdity of others' interpretations of his insights.

No such luck. The less sense Schrodinger's cat makes in our perceivable world, the more tantalizing it is to most science teachers as an introduction to the mysteries of the quantum world, and it is widely taught to explain the very concepts it was intended to disprove. The cat becomes alive or dead, in this experiment, only when we open the box and observe it. Until then, like denizens of the particle world, it is in limbo, neither one thing nor the other, with the potential for being either. Schrodinger's joke backfired.

Quite a few people are content to live in a universe that does not follow predetermined rules, in which events might as well result from a toss of the dice. They would certainly not sympathize with the physicist who, according to his friends, killed himself as quantum theory was developing because he could not bear the philosophical implications. His reasons to live, apparently, were based upon the conviction that life is reasonable and the universe works according to laws that are strictly observed.

Only a third of the way through the twentieth century, Hubble had shown that the still and placid night sky was a hurly-burly of huge galaxies speeding away from each other, as they had been doing for some billions of years, and Einstein showed that time and light were not at all what they seemed to be, and Bohr and his colleagues showed that the bedrock of creation is probability rather than certainty.

We may be more sanguine today. We are about to become familiar, it is predicted, with quantum computers. Instead of using transistors or chips to store bits of information, such devices would use electrons or neutrons or protons. In effect, the quantum, an indivisible unit of energy, would become synonymous with the bit, an indivisible unit of information. In theory, a single atom could be used as a computer. Quantum psychoanalysis (don't ask right now) is the subject of several recent publications. And to begin to attempt to understand the beginnings of the universe as we know it (see chapter 5), we have to imagine in some way the mysterious "quantum foam"—the cosmic birth medium. It is composed, perhaps, of bubbles of space-time, each of them approximately a millionth of a trillionth (!) the size of the nucleus of an atom. Each of them, theoretically, can give full glorious birth to a universe the size of our own—although no one in this universe would ever know about it.

We have long known that our thoughts are the daily business of organic computers. We are comfortable, I believe, with the idea that electrochemical reactions convey at very high speeds the information necessary to keep us in touch with our feelings, warn us of danger, move us to tears in the presence of great beauty, and shut us down for the operations of grief and despair. These communications happen before we can begin to perceive their effects.

But we now know that they are also being conveyed with bizarre imprecision and unpredictability at some level of physics. While our thoughts seem to be defined enough, most of the time, for handling actions and reactions in the world of consciousness, do the billions and billions of ambiguities of the subatomic world at the foundation of brain and body call into question our sense of self, our definitions of personality, our capacity for intentionality? Can we know much about who we are if our thoughts and emotions rise so outlandishly out of a quantum sea?

Later we will think about various approaches to the mystery of consciousness, including dualism. Getting ahead of ourselves just a bit, it's worth noting that the strange lessons of quantum mechanics have been invoked by Oxford University's Roger Penrose, among others, to explain how we individually, uniquely experience the world. The physical objects we see before us exist in the Newtonian world of classical mechanics in which, for example, no object can exist in two places or forms simultaneously. But our thoughts, our internal experiences, are by no means bounded by such restrictions.

But what if we consider the brain to be a quantum measuring device developed over millions of years of evolution as a link between the classical physical world and the quantum world? Consciousness, acting as Heisenberg's observer, would somehow cause (just exactly how, you will not be surprised, is the subject of intense theoretical debate) neural wave functions to collapse in response to sensory input. As H. Stapp has explained, "The mental life of each human being is representable as a sub-sequence of the full sequence of Heisenberg events." Some argue that the collapse of possibilities is random. Others propose that it must be in accordance with certain controlling principles lodged in the brain or, put another way, the individual self makes the choices. Theorists of quantum consciousness also have proposed that the brain's microtubules, a type of protein, are the physical structures that transform quantum states to classical states. The results, experienced by us in the rush of ongoing time, produce the constant thinking, imagining, and analyzing—the incessant evolution of personality and thought—that we handle as a matter of life's course.

And the continuing movement—the Earth revolving around our home star at 18 miles per second, our whole solar family whizzing around the black hole at the center of the Milky Way at some 200 miles per second, and the entire galaxy, as noted earlier, coming closer to our larger neighbor Andromeda at 50 miles per second—we handle that, too.

Besides, to keep from getting too dizzy, can we not look around at the comforting configurations of Earth with its towering stern alps and sturdy, deep fiords and ancient seas? Well, not really.

4

Wegener and the Dance of the Continents

Near the middle of the twentieth century there would come evidence of instability and constant change, if never chaos, at the very ground beneath our feet. Cliffs and mighty rivers, vast oceans and islands, have not existed in the same form and in the same place since the dawn of time. Every piece of the apparently stable globe is continually being churned up or pushed aside. What we see in any panorama is merely the most recent frame, another working sketch, the dough being kneaded. The present is not the end point, nor even a particularly significant way station.

We know all of this, but the basic concept of the "dance of the continents," which would eventually lead to today's continually evolving theory of plate tectonics, was resisted and ridiculed by geologists themselves for more than half a century. By the end of the nineteenth century, after all, they had discovered that Earth was at least hundreds of millions years older than anyone had guessed; one essential clue was the sluggish deposit of sediment over the centuries to create thin layers of rock. They did not find that this leisurely, deliberate construction of stable, mammoth earth structures jibed easily with Alfred Wegener's vision of do-si-do.

It is perhaps indicative of Wegener's affirmative attitude toward life that his controversial geological theories had

Alfred Wegener

nothing whatsoever to do with his fields of officially condoned expertise, meteorology and astronomy. The German scientist was as happy on the ski slopes as in his study (his favorite skiing partner and eventual son-in-law is the Brad Pitt role in the film *Seven Years in Tibet*), and although he believed his disputed theory correct, he did not continually play this one note in his life at the expense of others. He just had a fine idea that appealed—as it happened, in 1903 when he was 23 years old—and went gangbusters with it. Although nowhere near as conceptually breathtaking as Einstein's or Bohr's intuitions, Wegener's happy inspiration was right as rain, and a child could understand it, even as the experts hooted.

In fact, it is likely that millions of children seeing a globe or a map for the first time will notice (as did some of the cartographers who first began mapping the new latitudes appearing from the mists of European ignorance in the Age of Exploration) what Wegener noticed: with the right kind of earthmoving equipment, you could easily push all of Earth's continents together into one supercontinent. The jagged edges fit almost exactly.

But children grow up, and cartographers are meant to be receptors of information, not theorists. Practically no serious person a century ago—or so the absence of contrary evidence suggests—believed that Senegal and Brazil once embraced, or that unique Australia depended long ago from the belly of Africa.

Let us pause to be fair to all establishments. In the first place, anyone at a research center or a university or even a publishing house soon learns that there are a lot of fissionable cranks out there. A cursory look at Wegener's creds might have put him in that lockbox. Back in the 1970s an acquaintance, a math professor at an upstate New York college, lived in rank uneasiness if not terror because of a wandering lunatic grad student who occasionally bedeviled his days with the alleged proof of some famously unsolvable puzzle, like squaring the circle or finding the square root of -1. The young man was not only nuts but also violent. Eventually he disappeared or began taking his meds, but his kind are many if not legion: A certain kind of grandiosity is attracted to the unfathomable—and central to the affliction is the conviction that the keepers of the keys to conventional theories are either conspirators or fools. There are instances of over-the-transom revolutionary breakthroughs in science, but they are as few as supernovas.

Also let it be said for establishments, or for many of them, that they are often the achievements of thinkers who were once creative and cocky themselves. Young Turks too quickly and inevitably become the Old Guard, a large factor in the derisive reception given Wegener's ideas by the academy. Besides, they were human, perhaps agreeing with a character in Beryl Bainbridge's novel *Master George:* "Man himself is so buffeted by shifts of thought and mood, not knowing from one day to the next what he feels, that a shifting earth is well-nigh the last straw."

Finally, Wegener did indeed have a breathtaking idea, but little solid proof for it and no convincing theoretical explanation.

In the early twentieth century, meteorologists did not have chirping sensors spanning the globe, floating in the world's great oceans, or orbiting Earth. One effective way to understand weather was to go out and find it before it came to you. To learn from its extremes, you might journey to extreme places.

From Wegener's earliest expedition for the German government, a project to study Arctic weather conditions in Greenland in 1906, he began to notice odd physical echoes and resonances in widely separated places. Deep within the icy Spitsbergen islands, which lie above the Arctic Circle, great coal deposits had been found. But, as the academic geologists knew, they didn't belong there. Most coal, which is made from plant matter that has decomposed into peat and then been pressurized into layers of hard carbonaceous fuel, was formed 280 million to 345 million years ago in the Carboniferous Period. Essential to the process were huge amounts of vegetation and marshy lagoons or bogs. These ingredients were presumably not available so near the North Pole.

Wegener noticed other oddities of time and geography. If the tropical conditions for coal formation did not exist in the frozen North, the glacial action that forms vast, level plains of striated rock certainly never occurred in the hot, arid Karoo Desert in South Africa. Yet there stretched out exactly such natural pavements, a phenomenon that was typically found only in areas that had experienced an Ice Age, with retreating glaciers grading the rock to a flattened plane. Could climates have been differently distributed in the past? Not very likely. These puzzles evidently intrigued the very active Wegener even as he took up a post in 1909 at the University of Marburg, lecturing on meteorology and astronomy.

To glaciers in the desert and coal-nurturing steamy bogs beneath the ice, Wegener added the mystery of *Glossopteris* and the *mesosaurus,* which he had learned about by reading the work of professional geologists and paleontologists. The former, a fern that lived 250 million years ago, had been found in fossil beds in Australia, India, South Africa, and South America. Why would this single distinctive plant appear in such far-flung sites? *Mesosaurus,* a tiny aquatic reptile that lived some 300 million years ago, also was known from fossils found in a certain type of rock formation in only two places on the planet: southern Africa and eastern Brazil. Between these two sites lay 3,000 miles of South Atlantic ocean, surely an impossible journey for the fragile ancient swimmer.

By 1912 Wegener had formulated his basic theory of continental drift, as briefly explained in an article and lecture: the great landmasses had been joined together as one, a supercontinent he called Pangaea ("all land"), floating some 200 million or 300 million years before in Panthalassa ("all sea"). For unknown reasons, or no reason at all, Pangaea eventually broke in two, forming Laurasia and Gondwanaland. This was only the first step of the dance of the continents, which began to take their contemporary shapes and positions about 65 million years ago. Australia moved away from Africa, and the American continents shifted ever westward as the Atlantic rushed in to cut them off from Europe and Africa. Wegener began to work on a longer, more detailed exploration of his basic idea.

At this point—1914—humanity intervened.

A sincere pacifist, Wegener nevertheless signed up to fight for his country in the spectacular disaster known as the Great War. He was wounded twice, perhaps because he fought with the same vim and gusto he gave to everything else in his life. While recuperating or otherwise on leave, he wrote *The Origin of Continents and Oceans,* published in 1915, fully aware that it was only a sketch of a theory, with many important questions unanswered. In contrast with Darwin's tortured delay in completing and publishing the similarly startling *Origin of Species,* Wegener tended to rush into print, come what may, then rush back into print yet again when his precipitous explanations were discredited. This bad habit did not strengthen the case for his original accurate inspiration.

While it found some adherents, even within the academy, Wegener's idea was typically ridiculed. It didn't help that geology was dominated by British and American experts, or that many had unpleasant memories of the war. Far more important, however, was his continuing inability to offer an explanation, much less proof, although he tried on many fronts. What was the force, the engine that powered the dance of the continents? Casting about, he tried to enlist the tidal forces of the moon and the sun, as well as the rotation of Earth, but these were demonstrably much too weak to send mammoth landmasses skating away from each other. This was the "soft underbelly" of Wegener's idea, according to science historian William Glen: "He did not have a force by which to split Tangier and then propel its pieces across the face of the continent, across the face of the globe to take up their positions as the modern continents."

In the last months before his death at age 50 in 1930, Wegener began to hear from experts who were constructing strong arguments to support his overall idea, but there were still too many gaps. Stubbornly, he remained convinced that objections, however well founded, would eventually be discredited by new evidence. On his last expedition to ferociously misnamed Greenland, he was inspecting a new research outpost set up on the island's central ice cap when an early winter blizzard roared in. The station had only enough food for the two researchers assigned to work there. Wegener and his Inuit guide Rasmus Villumsen strode out into the driving snow to return to their main base. Neither survived.

Celebrated as a hero but not respected as a theorist, Wegener passed into obscurity. It would be another quarter century before unexpected discoveries beneath the sea, combined with observations of very peculiar evidence in rocky hills, would renovate and affirm his theory.

During Wegener's life, there simply wasn't enough knowledge to support any sweeping theory about the origin or nature of the Earth. He believed that mountains rose up when continents smashed into each other; others believed that the Earth was slowly cooling and shrinking, forcing up mountains or gouging valleys like wrinkles on a giant prune. A total of five different theories for mountain formation were taken seriously in Wegener's day. These were points of view rather than well-reasoned, well-founded scientific arguments.

Whatever was going on, at least experts agreed by the 1920s that it had been going on for a very long time. In the late nineteenth century Antoine Henri Becquerel, playing around with the newly discovered and wildly popular X rays, accidentally discovered radioactivity. By 1900 he understood that uranium was radioactive and that it loses its radioactivity at a steady rate, gradually turning to lead. A little more than a decade later, the technique of radiometric dating was well established: By measuring the ratio of the remaining radioactive uranium to lead in a rock sample, it is possible to estimate when the uranium was originally formed. By the 1920s geologists knew that our home planet is probably 4.4 billion years old. (Today's presumably more accurate estimate is some 4.6 billion years.)

Knowing the age of Earth, however, is not necessarily to know much about its history. Vast areas of the land above the seas had not yet been explored in the early twentieth century, and virtually nothing was known about the invisible ocean floor that accounts for some 70 percent of Earth's surface. Still, expert opinion agreed that the planet was covered by a thin crust, or rind, that included visible dry land and the hidden seafloor. A much thicker layer of hot, almost liquid rock and magma boiled just beneath the crust, according to the general opinion. This was known as the Earth's mantle. Occasionally the matter there spewed through the thin crust as volcanic lava. Below the mantle was some kind of core, its nature unknown and thus heartily debatable.

Presciently or luckily, one Benjamin Franklin had speculated about its nature a century before: "[This core] might be a fluid more dense . . . than any of the solids we are acquainted with; which therefore might swim in or upon that fluid." And he speculated further about the implications of this idea: "Thus the surface of the globe would be a shell, capable of being broken and disordered by the violent movements of the fluids on which it rested." What this may say about the true authorship of ideas is unclear, but it may underscore an important point about Wegener: it is not just that he was inspired by an idea but that he was determined to prove it. Often his tenacity led him to come up with half-baked "proofs" that did not bring a brighter gleam to his original proposition.

One problem with his image of drifting continents, according to the conventional schematic of Earth in his day, was that the thicker continental crust would have to break through the oceanic crust in order for the continents to dance away from each other. This action, something like an ice-breaking ship ramming through an Arctic ice pack, would certainly require the great force that neither Wegener nor any among his few adherents could identify.

Another problem arose from earthquake studies. Researchers knew that for several hours following a severe tremor, long shock waves vibrate completely around the surface of the globe, a tremendous wrench of force. Yet the continents do not dance, do not even budge a step, during this powerful vibration.

By the 1930s geologists may have felt themselves irrelevant or impotent in face of the exhilarating discoveries and theorizings of their counterparts in other fields: Hubble's vigorously expanding universe, Einstein's redefinition of gravity and portrait of space-time, Bohr's nutty subatomic particles. There was no similar buzz in the world of rocks.

That began to change when two men who did not believe in continental drift noticed unexpected peculiarities in the seafloor. Harry Hess, a Princeton University geologist, and F. Andries Vening Meinesz, a Dutch geophysicist, set out to determine the exact shape of Earth, known not to be a perfect sphere, by measuring the gravitational forces in different regions of the sea. On board a U.S. submarine, which could dive deep enough to escape wave action at the surface, they used a very delicate and sensitive special-built device, a gravimeter, to map the patterns of gravitational pull beneath the seas.

Their thousands of measurements, including differences as tiny as one one-millionth of Earth's gravitational force, dumbfounded them. For no reason they could conceive, large areas of surprisingly weak gravity were found along the Pacific rim.

And then again, humanity intervened.

During his service in the U.S. Navy after the events at Pearl Harbor, Hess commanded support ships in the Pacific Theater, often transporting fresh troops into the thick of battle. War begets technology, of course; newly improved sonar made it possible for Hess, when the military action cooled, to take echo soundings of the Pacific seafloor. He immediately discovered features beneath the waters that Wegener and his contemporaries had not imagined: Precipitous valleys were dug as deep as 7 miles below sea level on the ocean side of island chains that arced near the Asian continent. And these astonishing trenches closely paralleled the areas of weak gravity that Hess and Vening Meinesz had found with their gravimeter. Between forays that were key to the resolution of the greatest military conflict so far, the naval officer was forging the key to understanding the ongoing, much more massive forces that power the continually changing, writhing, shockingly fluid Earth.

Systematic mapping of the unknown seafloor speeded up after the war, spurred by military concerns about the nooks and crannies that might provide cover for enemy submarines or otherwise play a role in underwater combat. A mountain range with peaks 6,000 to 10,000 feet high was found in the mid-Atlantic. First noticed when the transatlantic cable was laid in the 1880s, this ridge had not previously been understood as echoing in shape the American and European coastlines. Soon it would be discovered to be only one of a series of ridges that join together to girdle the Earth underwater in a system 46,000 miles long. Often likened to the seam on a baseball, this great composite range was astonishingly active at many points. Volcanic islands such as Iceland and the Galapagos as well as submerged active volcanoes in the hundreds clustered along the ridges, a gathering that could not be serendipitous.

But the first discovery, named the Mid-Atlantic Ridge, was more than a dramatic underwater panorama. It immediately gave up a disturbing secret: aging of the rocks nearby showed it to be only 150 million years old. How could such a major geologic formation be so young on an Earth 4.6 billion years old? Later studies found that 200 million years was the upper limit in age at any part of the seafloor.

Hess, back in mufti at Princeton, considered this information and other undersea observations for some years. By 1960 he was ready to argue the case for a revolutionary idea known as "seafloor spreading." At this point the evidence was still patchy, but he had a gift for inspired guesswork; even so, he cautiously described his theory as "an essay in geopoetry." Hess summarized his ideas in a seminal paper, "History of Ocean Basins," published in 1962.

Each of the undersea ridges, he suggested, arose from within a great rift in the ocean floor, a crack in the oceanic crust. As in volcanic formation on dry land, hot magma churned up to form the peaks, pushing older rock to the side; this older rock had arrived on the ocean floor in the same way, but ages before. Therefore, as the observations showed, the rock in the central ridges would be younger than the rock to the side. Rising up slowly under the sea, the younger rock pushed the older rock away, causing the seafloor itself to spread at the rate of a few centimeters a year. In the Atlantic, therefore, the seafloor to the west of the ridge would be heading toward North America, the seafloor to the east toward Europe. Meanwhile, the differences in rock density on the ocean floor caused by the upsurge of younger rock would explain those tiny fluctuations in gravity detected by Hess and Vening Meinesz.

As the crust was pushed toward a continent, in Hess's geopoetry, it ran into a more substantial continental crust and was pushed underneath. This process he called subduction, a possible explanation for those deep trenches he had discovered between lulls in battle. "The continents do not plough through oceanic crust impelled by unknown forces," he proposed. "Rather, they ride passively on mantle material as it comes to the surface at the crest of the ridge and then moves laterally away from it." If he was correct, the ocean's crust would be forced to dive back down into the Earth's mantle, perhaps for geologic recycling. Subduction is not a negligible event. The Andes were formed when the oceanic crust was subducted along the continental margin of western South America.

The paper was hardly out of his typewriter before Hess, in an age when scientific discovery was accelerating much more quickly than in Wegener's day, was provided support in an unexpected way.

A nonscientist attending an exaltation of scientists is likely to be struck by the unprovable observation that, compared with their fellow beings, there is a fairly large contingent of pale young men who wear plastic pocket protectors for their pens, and an equally large contingent of hale and vivid young women and men who look as if they've just returned from several months atop Everest or in the jungles of Borneo.

In the latter category, presumably, was Allan Cox, a Berkeley geology student who was puzzled by a little-known but deeply unsettling mystery that had one of two possible answers: Either the North and South Poles have somehow changed places throughout geologic history, or magnetic information in certain rocks on Earth reverses from time to time. To solve the mystery, Cox went out into the foothills of the Sierra Nevada near Bishop, California, to retrieve samples of basalt, the rock formed from hardened lava.

Previously, geologists had learned that such rock, as it cooled into rigidity, became magnetized by the Earth's magnetic field so that its own magnetic field would point ever northward like a reliable compass. Why, then, would intrepid geologists on expedition to Iceland find basalt with a magnetic field pointing directly in the opposite direction, toward the South Pole? Other wrongheaded basalt compasses were found in Europe, North America, and Japan.

Cox, reviewing data that suggested that the age of a specific basaltic compass might have something to do with the direction of its magnetic field, decided to explore what this apparent connection might mean. If there was a pattern, perhaps the poles of Earth had indeed "flipped," bothersome as that might be to common sense.

In fact, Cox and his colleagues found that half their California basalt sampling pointed to the north, half to the south. It was not until 1966, however, that Cox and his friend Brent Dalrymple, an expert in dating rocks, were able to date the various flip-flops in the Earth's magnetic field: at irregular intervals, for reasons that have not been explained yet to anyone's satisfaction, the reversals have occurred nine times over the previous 4 million years. During these successive periods, which might be as brief as 100,000 years, an imaginary compass would point to one pole, then swing around to the other when the magnetic field oscillated. As will become clear, no human being has ever witnessed such a gut-wrenching event.

Before the pattern of reversals was established, however, two other eager researchers leapt upon the same data that intrigued Cox, but from a slightly different perspective. Using magnetometers trolled behind a research ship in the northwestern Indian Ocean, Fred Vine and Drummond Matthews discovered in 1961 a phenomenon on the seafloor that does not exist on dry land. Usually found parallel to the midocean ridges central to Hess's geopoetry, long miles-wide ribbons of black and white stripes, immediately dubbed "zebra stripes," alternated with each other on the ocean bottom.

Now it all began to come together, solidifying Wegener's unsubstantiated hypothesis into scientific certainty. Hess's lava flowing up to create a ridge would, like the California basalt, become magnetized in accord with Earth's magnetic field as it cooled in the chill deep salt sea during a particular epoch. The stripes, as they edged toward the continents, maintained a record of the primordial reversals of the Earth's magnetic pole. If this was indeed the case, the stripes should have at least roughly the same alternation of thick and thin lines on each side of a midocean ridge. Further research affirmed that this predicted symmetry was in place. In 1965 both Dalrymple's research team and an expedition of undersea experts discovered, virtually simultaneously, a previously unknown flip of the magnetic pole at 900,000 years before, the most recent such event. From Cox and Dalrymple's work to Vine and Matthews's work to other bits of information coming in, the evidence was persuasive: the flipping of the poles as registered on the seafloor proved that Hess was right.

And if Hess was right, so was Wegener. The widening oceans would have put pressure on the continents from the era of Pangaea, causing them to drift so that the Karoo became tropical and the Spitsbergen islands became Arctic, driven long and hard from their original spots on the globe.

Today, as the continents are steadily pushed away from the active midocean ridges—for example, as the Atlantic widens by about 4/10 inch every year—Wegener's and Hess's theories have been brought together in so-called *plate tectonics*. Earth's crust, probably no more than an average 60 miles thick (43 miles average beneath the oceans, 93 miles average under continental landmasses), turns out to be a patchwork of grinding plates, incessantly strong-arming each other for more room.

At the boundaries where they clash, earthquakes and volcanic activity occur. When two continental plates bang into each other, they can create a great mountain range like the Himalayas at the collision site of the Indian and Eurasian plates, just as Wegener himself once surmised. Where two plates slide against each other, they cause a transform fault at what is known as a strike-slip margin. There the Earth can quake and tremor with dynamic force, wreaking widespread destruction in a flash of edgy shearing. Along the infamous San Andreas transform fault, the Pacific Ocean plate is sliding inexorably northwest, shoving southern California toward its more decorous northern cousin by about 2 ⅓ inches a year.

There is still some disagreement about the exact boundaries of the plates, since we are observing them in a tiny fraction of geologic time and action. Perhaps there are thirteen major plates; the largest estimate is twenty. The average plate is not more than 1,860 miles wide, but the Pacific Ocean plate is about four times wider than that. Future dramatic events may more clearly delineate the margins. The Indian plate is slamming northward into central Asia, while Australia, nestled on the same plate, may be splitting off. As in so many other observational challenges brought to the fore by discoveries of the past hundred years, we are prisoners here of the end point fallacy, taking the world of our experience as a fact of completion rather than a continuing evolution. Today's Pacific Ocean and Madagascar and K-2 are no less or more transient in being than Pangaea and Panthalassa, or the shape of Ursa Major, or the average distribution of a particular type of DNA in *Homo sapiens,* or the size of the observable universe. There is no *terra firma,* no solid ground, except in the tiny pocket of time a short-lived sentient being thinks of as existence.

While continental drift was becoming accepted in the 1960s, geologists also were learning through experiments the likely structure of Earth from core to crust. Once again, military objectives played a major role in promoting the science that by accident would fill in even more of the blanks in Wegener's great theory.

This time it was not war, but rumors of war.

In order to enforce a nuclear test ban agreement during the Cold War, a series of very sophisticated seismic monitoring systems was set up around the Earth, adding to the thousands of seismographic stations already in place. Like the tremors and aftershocks of earthquakes, the low-frequency energy waves caused by nuclear explosions pulsate for long distances through the ground; the severity and the location of an event are easily detectable by seismologists with their seismographs.

These instruments also measure how seismic waves are reflected or bent in their journeys by the different densities of rock at different levels, thus producing a map of types of rock in the Earth's crust. At the same time, the waves are absorbed by liquids, so wherever the waves cannot be detected in the ground beneath, the area cannot be solid rock.

Even as this more complete picture of the entire crust and mantle was being formed, seismologists were able to make a snapshot of subduction in action along the margins of earthquake zones. Huge slabs of the rocky crust could be measured by the shock waves they produced diving back into the mantle, pushed there by seafloor spreading. The hot, partly molten mantle was shown to be some 1,740 miles deep, reaching down to the edge of Earth's core. (Drilling equipment has penetrated only about 7½ miles down.) From that deep verge up to the bottom of the crust, temperatures ranged from 6,690°F to about 2,000°F.

So far this depiction is a refinement rather than a refutation of the accepted view in Wegener's day: a thin, rigid crust including seafloor and dry land atop a mantle filled with flowing rock that becomes hotter and more liquid toward the center of the Earth. Today the crust and the top portion of the mantle with its relatively dense molten rock are together known as the lithosphere (from the Greek word for stone, *lithos*); the less dense rock in the mantle below is the asthenosphere (from the Greek *asthenes,* meaning weak).

Somewhat more is known about the core than in the 1920s, however. Geologists believe that it is divided in two: a hot liquid outer core, principally molten nickel and iron, that is more dense than the mantle above; and an inner core of solid iron and nickel, incredibly hot but kept solid by intense pressure. The radius of the swaddling outer core is about 1,860 miles.

Evidently the core itself is a giant furnace, but whether it produces powerful thermal currents that propel the upthrust of midocean ridges and how its action might affect the flip-flop of Earth's magnetic field are only two of many questions that have not yet been answered.

Back in 1928, only a couple of years before Wegener died, a geologist suggested that the intense heat of volcanism was caused by convection currents beneath the surface of the Earth. The same process occurs in a boiling pot of water: Water heated at the bottom of the pot rises up to the surface and out to its edges, cools slightly as it bubbles, then sinks back down as another current of heated water boils to the surface.

Wegener naturally seized upon this idea; it was a potential explanation of the engine for continental drift. Unfortunately, not enough was known about the Earth's mantle for anyone to speculate how convection currents could be shoving the continents away from each other.

In today's model of inner Earth, it is possible, but by no means proved, that convection currents rise very slowly from the outer core upward through the sludgy mantle—in fact, taking millions of years to reach the top—and then behave like the water boiling on the stove, cooling and sinking back toward the core. Some action similar to the turbulent surface of boiling water, though not detected yet as cause and effect, might be the missing motor of continental drift. But it would be infinitely more complex than the homely stovetop analogy, since the currents would be moving up and down through various types of rock and also be affected on their long, slow round trip by changes in pressure and temperature.

Powerful as the great engine of the Earth's core may be, implacable and impressive as the rise of mountains and the collision of whole continents are to us, the Earth is but a tiny part of the solar system, whose obscure star is one of an uncountable number of stars in the entire observable universe—literally more numerous, it is thought today, than the grains of sand on all of the beaches of this planet. Yet it all began, every massive star and galactic supercluster together, Mount Everest and the slim, shy Titicus River at the bottom of my drive, as a universe the size of a subatomic particle.

5

Big Bang, Big Crunch, and Big Bore

Science made this particular pope very merry.

"Thus with the concreteness that is characteristic of physical proofs, [science] has confirmed . . . when the cosmos came forth from the hands of the Creator. Hence, creation took place in time. Therefore, there is a Creator. Therefore, God exists!"

Actually, "science" confirmed no such thing. Pope Pius XII exulted in 1951 because of a new theory, not then definitely proved but rapidly gaining adherents among working secular cosmologists of the day, that pictured the beginnings of the universe as a specific instant. The big bang theory is no vindication of the Genesis account of Creation (besides, there are two separate accounts in chapter 1); it is simply Hubble's expanding universe reversed.

In other words, since galaxies are dashing away from each other at enormous speeds, is it not reasonable to suppose that they all somehow emerged from the same place, like stars from a rocket burst on the Fourth of July? In that case, all of the matter observable by us—all galaxies, celestial objects, nebulae—must have been crammed together very long ago as something unimaginably dense.

In addition, an expanding universe, according to Newtonian physics, would be slowly cooling down as it spread out, creating space; therefore, back in its beginnings, it must have been incredibly hot.

To sum up these two speculations, the origin of all things must have been a single point of infinite density and infinite temperature. Be warned that these ideas—infinite density, infinite temperature, vast reaches of time-space, and many more to follow—are not easily or finally grasped. Words can only do so much, only reach so far. Brain cells are presumably fixed atomic structures in a classically physical world, not quantum foam. Infinity is not readily accessible to finite consciousness. We are all in the same boat here, no matter how many clever analogies can be adduced to suggest inconceivable physical realities.

For example, take the little phrase slyly slipped in two paragraphs back: "creating space." If the universe is expanding, what is it expanding into? The standard answer is that the question is nonsense. The universe exists only where it exists; nothing else is. Therefore, expansion creates the space that defines the universe; it is not expanding into areas of something else. At some level, this concept is surely unimaginable to most human beings, but at another level, it is a concept that surely also delights the mind wrestling with it. And there's much, much more. "Asking what happens before the big bang," says Stephen Hawking, "is like asking for a point one mile north of the North Pole." Why? Because the linear, day-following-day time that enshackles us may not have existed before the dimensions that came into existence with the explosion; before then, time may have been a kind of circle. St. Augustine put it somewhat differently sixteen centuries ago: "The world and time had both one beginning. The world was made, not in time, but simultaneously with time."

But let's backtrack. Early in the twentieth century, more than one astronomer or mathematician recognized that

Einstein's theory of relativity predicted an expanding universe before Hubble's observations confirmed the idea. The "father of the big bang," the Jesuit priest and astronomer Georges-Henri Lemaître, suggested in the 1920s that a huge "cosmic egg" exploded: "The evolution of the world could be compared to a display of fireworks just ended—some few red wisps, ashes, and smoke." Accounting for our twentieth-century perspective, he added, "We see the slow fading of the suns and we try to recall the vanished brilliance of the origin of the worlds." As it would turn out, this cosmopoetry was a fine imaginative leap, although the details would be astonishingly more complex in all four dimensions.

But despite the apparent confirmation from Hubble's discovery, the next generation of cosmologists included several who were distinctly uncomfortable with the theory of cosmic birth. Their science included the basic tenet that the universe is homogeneous in every way, including time. In 1948 Thomas Gold proposed the steady-state theory: the universe would look the same to an imaginary observer placed in any part of it at any time throughout all time—past or present or future.

What about expansion? Gold and the many scientists who supported his alternative to the big bang theory believed that galaxies do speed off and eventually die off while new galaxies are born to take their place. Put another way, as matter raced away from other matter, new matter—theoretically in the form of hydrogen atoms—would be created to fill in the holes. The balance of matter would remain essentially stable, a steady state. (The creation of such new atoms had not, by the way, been detected.)

The colorful British astronomer Fred Hoyle rushed enthusiastically to the defense of Gold's theory, dismissing the big bang theory as "just not dignified, or elegant . . . rather like a party girl jumping out of a birthday cake." Ironically,

George Gamow

it was he who coined the term "big bang" in off-the-cuff remarks deriding the idea and was not amused that it immediately caught on with proponents and critics alike. Hoyle evidently felt a deeply personal antipathy to any sort of cosmic egg; he believed, and struggled to prove in several books, that the universe has "neither a beginning nor an end."

Steady-state theory could long endure, of course, only if no huge anomalies were found in the structure of the universe. By definition, it had to be basically the same everywhere and look the same from every imaginable point of view. To prove that was the case, however, involved proof of a negative.

By contrast, big bang supporter George Gamow suggested in 1948 that there was the possibility of positive proof for the primordial explosion: The heat and light of such a (literally) cosmic event would have spread to all regions of the continually expanding universe, slowly cooling down over billions of years. The result would be a kind of cosmic background radiation. Gamow predicted that it would have moved down the electromagnetic spectrum to become microwave radiation of about 5° above absolute zero on the Kelvin scale, the temperature at which all molecular action ceases. That would be −450°F, or −268°C.

Then came the unnerving discovery in 1963 of the most luminous objects in the universe. Billions of light-years away from us, these mysterious sources of radio waves, which are strongly evident in radio telescopy, are only dimly detectable by the most advanced optical telescopes, but their average actual brightness exceeds that of 300 billion suns. And that's just the average. Some standouts give off more light at their source than 30 trillion suns. One has been found to have the luminosity of 1.5 quadrillion suns. So luminous, so distant, they were christened "quasi-stellar radio sources," or quasars. According to the redshifts in their absorption line spectra, they are moving away from us at tremendous velocities.

So much for a homogeneous universe throughout all time, as the steady-state theorists proposed. These monstrous beacons were not only far away and ancient, they also were found only in other quarters of the universe. None of them was anywhere near us. The imaginary observer noted before would get a very different view of the universe than anything we can see from Earth if set in the time-and-space neighborhood of a quasar. Billions of years ago, the observer could have seen young or newly forming quasars; today we cannot. In sum, the universe turned out to have a history. Even if the quasar redshifts were misleading and the objects are closer than calculated, as some scientists have suggested, it is still clear that steady-state theory comes to grief in their bizarre, bright glare.

That was the proof of a negative to discredit steady-state theory. In a homely tale that no one can tell without mentioning bird droppings, there came positive proof in 1964 to support the big bang.

Radio astronomy grew by leaps and bounds in the years immediately after World War II, detecting the first pulsars, mapping the Milky Way's spiral galactic structure, and piercing to the heart of galaxies whose centers are hidden by clouds and cosmic dust. At Bell Labs in Holmdel, New Jersey, where government subsidies flowed in response to the space race, two young radio astronomers, Robert Wilson and Arno Penzias, were trying to get a clear reading of radio emission from a supernova, but they were peeved by a continual interference.

Their equipment, designed for satellite communications, had a 20-foot-long antenna that would soon become one of the most famous devices in the history of science. It could not detect the source of this persistent distraction. The static came from all directions rather than from an identifiable source on Earth or a single celestial object. In their own words later, it was "within the limits of our observations, isotropic, unpolarized, and free from seasonal variations." Frustrated, they resorted to spritzing the interior of the horn antenna to remove residue left by a flock of roosting pigeons. The interference was unaffected.

Here the gods become playful. Penzias and Wilson, just trying to get some useful work done, were not trying to probe into the beginnings of time and space; in fact, neither believed that the big bang made much sense. Not far away, however, physicists at Princeton University, believing very much in the theory, were trying to find the background radiation that Gamow had predicted. (In another ironic twist, perhaps, they did not even know about his prediction, having come to a similar conclusion on their own.)

The leader of the Princeton team, Robert Dicke, and his colleague Jim Peebles agreed that any cosmic background radiation was likely to have cooled down to the microwave spectrum. But just as they began cobbling together a makeshift radio antenna with parts gleaned from surplus stores in Philadelphia, word of mouth alerted them that the frustrated Penzias and Wilson had measured their interference as 3.1° above absolute zero on the Kelvin scale. Dicke wryly announced this news to his colleagues: "Well, boys, we've been scooped." Indeed, the Holmdel hiss had led to one of the scientific stories of the century: proof that the big bang theory is correct in its basic assumptions.

As we will see later, the background cosmic radiation, which has been more accurately measured today as 2.73° above absolute zero, does not spring from the first instant of the birth of the universe. It arose some 300,000 years later, marking the onset of the so-called Decoupling Era, or 12 billion to 20 billion years ago. To better than one part in ten thousand, it is uniform (the isotrophy that Penzias and Wilson mentioned) throughout the universe; that means it originated in every part of the existing universe at the same time. Also, it has a spectrum that indicates it began in a state of perfect equilibrium. In other words, when it was created, there was a balance between all of the matter in existence and all of the radiation.

In popular science writing these days, we are warned not to picture the big bang as a continuing explosion, despite the incredible primordial blast, but as an inflating balloon. All matter as we know it, in this image, lies on the surface of that bloating sphere. (The image does not perfectly suffice as an analogy; the "surface" of this model of the universe is actually three-dimensional.) Another image used frequently is the baking loaf of raisin bread; as the loaf grows larger, the raisins, representing galaxies, move away from each other, as if mutually repelled.

This caveat—expansion, not continual explosion—might seem to be a distinction without a difference at first hearing, but it helpfully reminds us that the growth of the universe is apparently orderly overall, if perhaps not evidently so in the maws of black holes and ferment of galaxies in the throes of birth or death.

As the universe expands and the galaxies move away from each other, as we expect, they do not expand within themselves; gravity holds them together in their spirals, ellipses, or other group configurations. This is the huge galactic dance of immense celestial structures. It began as something smaller than an atom, and its birth throes have to be measured in infinitesimal fractions of a second.

A careful reader might note that this is all so unbelievably crazy it makes biblical Creationism seem scientific.

Well, let's see. In the first place, cosmologists disagree about what could, did, or could not exist before the instant of the primordial explosion. The issue may not be resolved soon. Could the universe arise from nothingness, defying smug old King Lear's quip that "nothing will come of nothing"? He was talking about a dimple in the space-time continuum, inheritance of part of an ancient kingdom, but the idea may hold for space-time itself and for all of its physical laws.

On the other hand, even if matter did not exist, at least as we know it, were the ingredients popping in and out of existence in some sort of quantum soup, waiting for the right spark to energize the cosmic egg? Even in a pre–big bang theoretical vacuum, the tiniest of preatomic particles might have existed or enjoyed the possibility of existence.

But we don't know.

What we may know, as of this writing, is something of the timeline, broken down into quanta of inconceivability. Each of the following instants, these fractions of creation, is about as astonishing as anything you will ever read.

The universe began with the Planck instant, so called to honor the physicist Max Planck. Also known as the singularity, because the laws of physics as we know them did not exist, it lasted for 0.0000000000000000000000000000000000000000001 second, or 10^{-43} second. Possibly, if string theorists are right (and there are five competing theories), as many as ten to twenty-six dimensions existed just before the singularity, but they were trimmed down to the four space-time dimensions of our universe within the Planck instant.

What else was going on? Was this brief time chaotic, as the first physical characteristics of our universe were hammered into existence, or was the process orderly? We can't know exactly because there is an apparently impermeable barrier between our universe and the world of the Planck instant. It is thought that the four physical forces that govern today's universe, however, did exist during the Planck instant: gravity, the electromagnetic force, the strong force, and the weak force. They were combined in one unified state.

At the end of 10^{-43} second, the singularity ended, Planck time began, and the universe was created. It was a sphere perhaps as small as a proton, or perhaps as wide as the point of a needle (0.0039 inch). The matter that forms everything detectable in the known universe, all of the 50 billion galaxies found to date across the 13 billion or so light-years of observable objects, was packed into this incredibly dense sphere during Planck time. Roiling around in a kind of opaque "soup" that formed the cosmic egg were elementary subatomic particles like electrons and neutrinos; atoms did not yet exist. Before that could happen, the four then unified forces would have to break apart. For now, only gravity became distinct from the other three forces, which would stay together only until the next epoch, known as inflation.

Inflation, even stranger than everything else so far, apparently happened when Planck time ended with the universe 10^{-36} second old: According to the theory of inflation, not added to the big bang theory until 1980 by particle theorist Alan Guth, the trillions of atomic particles within the cosmic egg suddenly raced out at the end of Planck time from a volume of nearly zero to spread almost evenly throughout the universe. In this split second of inflation, evidently caused by a minuscule change in energy, the universe almost instantaneously grew billions of times larger, at the same time releasing oceans of energy that created more matter. All matter, whether detectable yet by human instruments or not, was now in existence. The three weak forces disbanded. Eventually the electromagnetic force would act to hold atoms together in their characteristic structures. The strong force would come to hold the nuclei of atoms together. Certain kinds of radioactive decay in atomic nuclei would be governed by the weak force. Guth's theory has been accepted for now as the best existing answer to a puzzle: What kind of expansion could have resulted in a universe that looks to be relatively homogeneous, with galaxies evenly distributed and an even temperature throughout? During inflation, all of the matter existing during Planck time suddenly spread evenly to all parts of the expanding universe; hence the continuing homogeneity—defined as evenly distributed temperature and galactic structures—that prevails overall. Everything sprang from the same place at the same time.

Between the end of inflation at about 10^{-33} second and the end of the first second of the big bang event, the predominant matter was in the form of heavy elementary particles called hadrons, which include all of the types of subatomic particles that are formed from quarks. The temperature of the universe cooled down from infinite temperature to about 3 trillion degrees Kelvin, making it possible for quarks to stick together to form neutrons and protons, which had been impossible at higher temperatures. Other combinations of quarks that were too complicated to survive were annihilated, knocked apart by the heat before they could form.

As the universe cooled, its organic processes began to slow, comparatively speaking. Its second second of existence began the leptonic era.

This era began as the universal temperature dropped to 10 billion degrees Kelvin, thus allowing the domination of the lighter atomic particles known as leptons: neutrinos, antineutrons, electrons, and positrons.

Some 60 seconds or so after big bang, the nucleosynthetic era was inaugurated, with temperature dropping to 1 billion degrees Kelvin, and continued for about 2 minutes. At this point the entire universe resembled in heat and density the center of a young star, where thermonuclear fusion continually creates helium. In these conditions during the early big bang, helium nuclei were produced in huge amounts in a three-step process. A nucleus of deuterium was created when a proton was captured by a neutron; the deuterium changed itself into tritium by bagging another neutron; and the tritium caught another proton to make itself into helium.

If you've been counting, you note that a helium nucleus is made from two protons and two neutrons. While the helium nuclei became by far the most numerous, very small numbers of deuterium and lithium nuclei also were stabilized by nucleosynthesis; the ratios among these three elements, to be formed later when atomic structure became possible, have remained the same as the universe has expanded for its 12 billion to 20 billion years of existence.

Helium accounts for about a quarter of the mass of the universe. The production of helium nuclei during the intensely hot first moments of the big bang was predicted by theorists such as Dicke, since such conditions would be necessary to explain its relative abundance. The only other possibility—manufacture in the furnaces of stars—would produce comparatively small amounts of the element.

At this point, 3 minutes have passed.

When the frantic thermonuclear activity that created helium ceased, high-energy gamma rays and other radiation dominated the universe. This was the beginning of the radiation era, which would continue for some 300,000 years. For the first 10,000 years a mammoth fireball burned at the center of everything. Slowly, the amount of energy stored in matter began to overtake the amount of energy stored in radiation, a crucially important step toward the creation of atoms. Photons, the tiny quantum particles of light, could not escape and travel outward during this period. Free-roaming electrons kept knocking them around, perhaps millions or billions of times, in a phenomenon known as scattering. In other words, if our imaginary observer happened to be in the neighborhood but outside the universe, he or she would not be able to see any light beaming outward; during its first 300,000 years of infancy, the activities of our universe were shrouded from view. This is only one of the reasons that many of the artists' renderings meant to represent the big bang probably have more to do with inhaling than with the minutiae of the first flowering instants of existence.

The fog lifted when the universe grew to about 1 million light-years (say, 60 billion billion miles) in diameter and cooled to 4000° K. That was temperate enough to allow subatomic particles to combine with each other to form hydrogen atoms, a process called recombination, which then coexisted with the helium formed at the previously higher temperatures. Since electrons were almost all trapped into orbit around atomic nuclei, light and other radiation were no longer scattered. Decoupling, as it is called, meant that the radiation moved independently of matter and raced away, thus creating a visible universe, a mammoth cloud of transparent gas.

If Yahweh ever said "Let there be light," this was the time.

For the next million years, recombination continued so efficiently that, at the conclusion of the era, only about one electron and one proton were still roaming free in comparison with every hundred thousand atoms. Meanwhile, the photon radiation continued apace, its temperature dropping toward the very low temperature discovered by Wilson and Penzias with their horn antenna.

Now comes a huge leap. Once matter was formed and hydrogen gas continued expanding outward, the stage was set for the birth of stars and galaxies. Much is still unknown, but current speculation holds that the process of galaxy formation probably began 1 billion to 2 billion years after the Planck instant. Perhaps it took another billion years for the first youthful stars to stabilize. For the next billions of years, some of these stars would become blast furnaces for the formation of the heavy elements that are familiar and essential to our lives on Earth: iron, oxygen, and carbon, among others. When one of these stars exploded in the dramatic death of a supernova, the elements were spewed out across the cosmos. Eventually they would become critical to the construction of planets in our solar system, which began forming about 10 billion years after the primordial explosion, and would play their roles in the composition of our bodies and all other known life. It was the late Carl Sagan who first pointed out that thus we are creatures made from stardust, an observation that bears repetition: the child's hand that reaches out to touch yours could not exist without the preparation we have just rehearsed from Planck time through the radiation era to galaxy formation to the continually evolving present. The most mundane furniture of our lives, the beauty and the gimcrackery, are made from the detritus of supernovas millions of light-years from us in space-time.

And yes, our own Sun will die relatively soon, probably in about 5 billion years, but not as a supernova. It is too small, too average, to explode across light-years of space and hurl heavy, life-building elements toward waiting worlds in the process of formation. It will simply run out of hydrogen and end up losing everything but its Earth-size core, cooling to become a so-called white dwarf, then finally a black dwarf, joining billions of other stars in the Milky Way that have already suffered the same fate.

Will the big bang theory hold up long? Most cosmologists think so, although they also know from the experience of the twentieth century that hugely transformative discoveries can occur overnight. For the four decades it has prevailed, however, some troubling questions or objections have been given dramatic affirmation in experiments.

For example, the expanding universe is homogeneous overall, but it does have great structural diversity. The most astonishing example is the Great Wall, a gargantuan sheet of galaxies that is more than 500 million light-years wide. No other structure nearly so breathtakingly large has yet been found. At the same time, we now know that bubblelike voids some 150 million light-years wide can be found between galactic superclusters. We have evidence of diversity much closer to home as well. Our own galaxy lies within the Local Group, where the density of mass is some two hundred times the average mass density throughout the cosmos.

Gravity, as suggested earlier in the balloon and raisin analogies, is responsible for holding huge galactic structures together. Where there is a so-called nonuniformity of matter, gravity begins a process of intensifying the nonuniformity by dragging more matter into the area, then more matter, and so forth, as the structure grows. This gravitational clumping, however, begins only when a nonuniformity exists to begin with.

What would produce these phenomena?

If we suppose that the expansion of the universe increased exponentially, as happened during inflation, it would be filled with intensely hot radiation, a state of affairs in which hot matter would be banging around in aleatory fashion. Such thermal fluctuations, as they are called, could very well cause density perturbations—that is, the nonuniformities that attract gravity and are turned into discrete structures. In fact, cosmologists have calculated the magnitude of density perturbations at recombination, or 300,000 years after the big bang.

The proof? Well, you have probably noted already in this chapter that unlike some theoretical leaps in science in the twentieth century, many aspects of big bang theory have been shown to be susceptible to experimental proof. In this instance the density perturbations—the seeds of future galaxies—would be recorded as thermal fluctuations in the cosmic background radiation.

And so they were. "Broad wrinkles in the fabric of space," as a *New York Times* science reporter phrased it, were indeed discovered in 1992 by COBE, the Cosmic Background Explorer. A satellite poised about 550 miles above us, COBE made some 210 million measurements with three separate microwave radiometers, or two complete surveys of the entire celestial globe over a 12-month period of observation. Scientific analysis revealed a phenomenon called anisotropy (as opposed to isotrophy), or irregularities in the density of matter as it was being formed 300,000 years ago. The "broad wrinkles" actually never varied more than 0.00001 degree.

COBE found that temperature variations in the cosmic background radiation today matched the structure of the universe: where temperatures were about 0.001 percent warmer than the average, galaxies and galactic clusters and superclusters glow brightly; and where temperatures are cooler than average by 0.001 percent sit the dark, bubble-like voids of space. These tiny variations in temperature were thought to show that acoustic waves roiled around in the infant universe in its first billionth of a trillionth of a trillionth of a second, producing the irregularities that seeded cosmic structures.

In 2000 the BOOMERanG experiment in Antarctica, which involved sending equipment in the gondola of a huge balloon some 23 miles into the stratosphere over the South Pole, produced data about thirty-five times as richly detailed as the COBE radiation maps. In an area covering about 2.5 percent of the sky around us, BOOMERanG found hundreds of the slight fluctuations in the cosmic microwave background radiation, confirming big bang theory.

Then in 2001 the first direct evidence of the waves may have been photographed through a telescope by an international galaxy mapping project called the 2-Degree Field Galaxy Redshift Survey (2dF). According to researchers analyzing the data collected by the Anglo-Australian Telescope outside of Coonabarabran, Australia, faint imprints of the primordial acoustic waves were as large as 300 million to 1.5 billion light-years, or approximately in the size range predicted by big bang theorists.

As galactic mapping continues with the five-year-long Sloan Digital Sky Survey, an $80 billion project set up to fix the positions of one million galaxies on a three-dimensional map, it seems probable that further physical evidence of galactic seeds will be found, spurred by the 2dF findings. At the end of its first year in mid-2001, the survey had already photographed galaxies more than 3 billion light-years from Earth.

A beginning implies an ending—usually. Cosmologists still debate whether the universe will end and if it does, how that will happen. Will the balloon expand until the energy pumping it outward is exhausted, then implode backward upon itself? That theory is known, of course, as the big crunch. Expansion runs out of gas eventually; all matter sinks back toward singularity. The mass and the energy combine to cause the collapse.

The future of the universe could be reliably predicted if we knew the relation of kinetic energy to maximum potential energy. Energy can be neither created nor destroyed, but it can, of course, be changed into its equivalent in mass or from one form of energy into another. Kinetic energy is the energy of a body in motion, like an arrow shot into the air or falling toward the ground. Gravitational potential energy, for that arrow, would be its potential energy at the height of its flight arc; at that point, still for a moment, it has no kinetic energy, but its potential is defined by its position in Earth's gravitational field. (If the arrow hits you on its way down, you'll know exactly how much potential it had at the top of its flight path.)

If the potential density of the universe is greater than its kinetic energy, the big crunch is to be expected; the universe will collapse back into a singularity. If kinetic energy is greater, this universe will expand forever. If the two forms of energy exactly balance each other at any time, the universe is said to have a critical density. If the actual density today is less than the critical density, expansion continues for tens of millions of years until eternity is dark and cold; if the density today is greater than the critical density, the universe will enter its collapse phase a few billion years into the future.

One disputed factor is the ghostly neutrino. An electron weighs sixty thousand times more than one of these scarcely detectable particles, but there are more of them in the universe than any other type of atomic building block—electrons, photons, or quarks. Some scientists believe that their collective mass could reverse expansion.

More likely, it seems, the universe will both expand forever and die out, a theory inevitably called the big bore. In this model, mass and energy are perfectly balanced to form a so-called critical mass. All matter in the universe veers away from all other matter into infinity. All of the stars become dark cinders, no new stellar nurseries are formed as hydrogen disperses, and atomic structure collapses. Something like 10,000 trillion trillion trillion trillion trillion trillion trillion trillion years down the journey of expansion, all that will be left is a vast, dark soup of loose electrons, positrons, neutrons, and radiation. What began as infinite density and infinite temperature will go on infinitely as cold, inert subatomic particles, their combined movement outward still creating space where nothingness had been before.

But the question is still open, as are others raised by big bang theory. Currently, the most ferocious debate in physics concerns the "age crisis." The annoying range of estimates for the occurrence of big bang, which you will find as 10 billion to 20 billion years at the extremes, results from disagreements about the rate at which the universe is expanding. Observations have not been able to nail down the exact distances of certain celestial objects; physicists disagree about the interpretation of redshift data for galaxies and other phenomena.

Einstein's cosmological constant complicates calculations since, much as he proposed, some sort of inexplicable force, dubbed dark energy, seems to be fueling expansion, causing matter to move outward faster than the balance between detectable matter and energy would suggest. A Hubble Space Telescope image of a supernova from some 11 billion years ago was interpreted in 2001 to prove the existence of this repulsive gravity, or dark energy. When this star exploded so recently after the big bang, gravity was stronger than dark energy throughout the toddler universe. Then, perhaps only a few billion years ago, the situation reversed: dark energy/negative gravity became dominant over gravity. This negative gravity, which has been estimated to account for 65 percent of the undetectable universe, caused the expansion rate to accelerate. The supernova's unusual brightness indicates that it was nearer Earth when it exploded—that is, when gravity dominated—than it would have been if it had exploded after the reversal.

Less debatable, but still mysterious, is the exact composition of subatomic material. Big bang theory predicts that protons and neutrons are composed of fundamental subatomic particles known as quarks, but they have not been directly detected. So many types of quarks and other fundamental particles have been theorized that some physicists believe that the apparent complexity results from misunderstanding or lack of evidence—that is, the most appealing answer to the fundamental nature of matter ought to be as simple as the best answers to other puzzles in physics. Simple but not too simple, as Einstein suggested.

To answer the debate, particle accelerators have been built to simulate the very earliest stages of big bang. Protons and antiprotons are propelled around huge tunnels in opposite directions until they reach 99.99995 percent of the speed of light—at which point their mass has increased, according to relativity, by a factor of several million—and then smashed into each other. In an explosion that lasts for only a trillionth of a trillionth of a second and reaches temperatures four hundred million times hotter than the interior of the Sun, the particles can fall apart, reverting under conditions that recall the earliest moments of big bang to the constituent fundamental particles that might have been created in just those conditions aeons ago.

Such experiments by their very nature verify the accepted theoretical model for the first seconds of existence. In 2005 the $4 billion Large Hadron Collider in Geneva will begin operation as the world's most powerful particle accelerator, drawing upon funding and scientific analysis from thirty-four nations. The device will use electromagnets to hurl protons in a stream around an 18-mile-long circle, causing them to collide at energies of 14 trillion electron volts. Over the last quarter of the twentieth century, previous colliders discovered evidence of charmed quarks, the tau, the gluon, the top quark, and other predicted particles. Now the race is on to detect a missing particle known as the Higgs boson. If found, it would prove the existence of the Higgs field, an otherwise undetectable field of energy that theoretically permeates the universe and provides the mass and the weight for all subatomic particles—and therefore for the entire universe—by interacting with them with a kind of friction. Otherwise, according to today's accepted theory, the particles would have no mass.

Scientists thought they had found traces of the elusive Higgs boson in 2000, but later analysis suggested that the supposed evidence was, as *New York Times* reporter James Glanz wrote, "a statistical mirage . . . random superpositions of ordinary particles." The suspense of this search may not be what poet Muriel Rukeyser meant by writing "The universe is made of stories, not atoms," but it is already high human drama marked by frenzied competition among research teams, conflicts over interpreting data, allocations of torrents of money, and aspirations to earn perhaps the greatest scientific prize of the twenty-first century.

Can it be noted too often? The atoms that are the building blocks of everything we see and otherwise experience, every thought carried electrochemically through our brains, every laugh or curse that is heard because of molecules of air, all exist because of a cosmic manufacturing process of truly inconceivable age, length, power, breadth, and, if you like, majesty.

The big bang is a far cry from the tick-tock, grinding little universe of only two centuries ago, but to some cosmologists there are even stranger, more unsettling possibilities.

If one cosmic egg was generated out of a primordial vacuum that seethed with potential particles, why not other cosmic eggs? Are there other universes expanding even now to create their own versions of space-time? Are they similar to ours, or did their birth create a different set of physical laws? Back to ideas of infinity: Is there any reason why there would not be an infinite number of such universes? Are big bangs as common as stars in our own universe?

6

Fermat, Godel, and Fuzzy Math

Compared with the theoretical uncertainties of astrophysics, the world of numbers is generally considered serene, concrete, infallible, immutable. One plus 1 always equal 2. There may be different types of geometry, but each follows the rules, however counterintuitive, that define it. Parallel lines never meet, as Euclid's theorems asserted to us in high school, or they do, in internally consistent geometries created in the nineteenth century. And the most famous mathematical discovery of the twentieth century, the proof of Fermat's last theorem in the branch of mathematics known as number theory, offers just that kind of comforting certainty. If the proof is correct, as is now thought to be the case, its correctness will not be affected by outside circumstances or new discoveries. But Fermat's last theorem, despite worldwide coverage, was not the signature math revelation of the past hundred years. The more important discovery was not only disturbing; it is also not widely acknowledged and understood. But more on that later.

Rarely does a Broadway review in the *New York Times* describe succinctly, or in any other way, a major intellectual discovery:

" . . . a proof of Fermat's last theorem, a mathematical conundrum postulated in the early 17th century by Pierre de Fermat, which stated that the equation $x^n + y^n = z^n$ has no solution when *x, y,* and *z* are positive whole numbers and *n* is a whole number greater than 2. Proving that this was so had stumped mathematicians for 360 years."

So wrote drama reviewer Wilborn Hampton in December 2000, after seeing a "chamber musical" called *Fermat's Last Tango,* evidently a far cry from *Cats* and *Fosse.* That such a work would be created for a popular audience and explained so blithely in a middlebrow journal is testimony to the wide attention given to the unexpected solution, late in the twentieth century, of a theorem that for a very long time was thought by many experts to be unprovable, and probably not all that much worth proving.

The centuries-long saga involves a revolving cast of characters more vividly individual than any Hampton saw on opening night. The real-life denouement finally jelled only after following the standard Hollywood three-act model: scientist finds solution, scientist loses solution, scientist proves solution. Phrases like "more elusive than the solution to Fermat's last theorem" appear in occasional pieces written for lay readers.

For those reasons it is worth recapping the Fermat puzzle, although its resolution does not lead to sweeping reevaluations of the nature of mathematics itself. The proof, though extremely long, complex, and based upon discoveries made in the twentieth century, does not challenge what has always been thought; it affirms it.

Pierre de Fermat, who lived from 1601 to 1665, was a deeply serious hobbyist. The father of five, he was a parliamentary judge in Toulouose, France, known to be conscientious and principled. It was in his spare time, however, that this famous Prince of Amateurs, as he became known, focused his remarkable energies and intellect upon the mathematical discoveries of ancient Greece, worked out the basics of calculus before Newton was born, and indeed founded the contemporary version of that branch of mathematics known as number theory.

From time to time he evidently enjoyed challenging other mathematicians of the seventeenth century with teasers that were so difficult that some experts, especially across the Channel in England, believed he was blowing smoke. That charge would be brought up unprofitably during the long search for a solution to his so-called last theorem.

In fact, as far as is known, this "theorem" was never broached to anyone else. After briefly describing it in the margin of a Latin translation of an ancient Greek text he was perusing in 1637 or so, Fermat wrote the comment that became one of the most celebrated lines in the history of mathematics: "a truly marvelous proof that the margin is unfortunately too small to contain." His proof, if indeed it existed, was never found. His so-called theorem, whether inspiration or vaporlock, would have been lost as well if his son, Clement-Samuel, had not happened upon the marginal notation and included it among his father's posthumous publications in 1670.

To enter the fray, we start with the Pythagorean theorem you may recall from high school plane geometry or, in certain lines of work, you use every day. Perhaps building upon numerical relationships found by the ancient Babylonians as long as 5,000 years ago, the Greek philosopher Pythagoras developed his theorem in the sixth century BCE: to wit, $a^2 + b^2 = c^2$ (a, b, and c are lengths of the sides of a right triangle).

His work, unlike the speculations of physicists and astronomers and geologists and others before the destabilizing discoveries of the twentieth century, is as eternal as it gets. As far as anyone can prove, even now, $a^2 + b^2$ will always $= c^2$. The same idea, also called a Pythagorean triple, can be geometrically expressed: The sum of the squares of the sides of a right triangle equals the square of the hypotenuse. If you've forgotten the terms, you'll remember a standard solution: When the sides of a right triangle are 3 and 4, the hypotenuse must be 5, because 3×3 (9) + 4×4 (16) = 5×5 (25).

While there are different kinds of geometry with differing laws, as we shall see, the numerical basis of Pythagoras's theorem does not change. In number theory, no matter how fast the spaceship is flying or how unpredictable the movement of electrons in the brain studying the problem, it is mathematically true apart from any physical considerations that $a^2 + b^2 = c^2$, as long as the numbers in question are positive integers, also known as natural numbers.

Glancing a moment upon this happy theorem for the nth time undoubtedly, Fermat evidently asked himself a pretty obvious question: If the relationship was true for squares, would it also be true for cubes and/or for higher powers? If and when would it be true that $a^9 + b^9 = c^9$, for example? Put another way, would there be Pythagoras-style theorems for any power in the form $a^n + b^n = c^n$ when n is greater than 2?

If we can believe the judge, he almost instantly answered himself *"Jamais!"* Never. No way. No other positive integer could work as the power in a Pythagorean triple, not from 3 to infinity. Why not? Because of that "truly marvelous proof"!

Talk about proving a negative. Over the next decades, a succession of gifted thinkers basically proved that specific groups of numbers would not work as "*n*." They could not, however, come up with a theorem that covered all possible positive whole numbers.

The club grew to have an impressive membership. Fermat himself showed that 4 was impossible in the equation. The Swiss mathematician, theologian, astronomer, and medical researcher Leonhard Euler proved in the early eighteenth century that the number 3 would never work as *n*. The powers 5 and 7 fell by the end of the century. Early in the nineteenth century the young Frenchwoman Sophie Germain, though forced by the mores of the time to disguise herself under a male pseudonym when corresponding with other mathematicians, was able to prove that a large class of positive integers under 100 could not be *n*. (Her life and career have been resurrected quite powerfully in a dramatic high point in David Auburn's award-winning Broadway play *Proof,* for the heroine needs Germain as inspiration, even at this late date, to believe in her own mathematical genius.) Following Germain's lead, a German expert, Ernst Eduard Kummer, proved that Fermat was right for all prime numbers under 100 except for three: 37, 59, and 67.

But if a trend might be seen to be developing here, no one could yet affirm what Fermat himself supposedly glimpsed in an instant: a general proof that the *n* in a Pythagorean triple could never be any other positive whole number but the number 2.

At age 10, according to the English mathematician Andrew Wiles's recollection, he first encountered Fermat's challenge in a book picked up by chance at his hometown library. "It seemed so simple. [The book] said that nobody has ever found a proof of this for over 300 years. I wanted to prove it." You can see his point. If 2 is only one of an infinite array of positive integers, how likely would it be that it alone could play a role in a Pythagorean triple?

Reality, in the guise of academic fashion, distracted him for years; for a student concentrating in math to attack Fermat's last theorem would be as sensible as an aspiring astrophysicist announcing an effort to prove that galactic redshift is an illusion (a subject addressed today by contrarian experts who already have a degree). Evidently a practical human being, Wiles earned his doctorate in number theory at Cambridge on an acceptable subject, elliptic curves, and was hired by Princeton University with the idea that he would continue working on that and related topics.

Wiles did not, at least at first, need the advice a concerned mathematician father—Hungarian Farkas Bolyai (1775–1856)—once gave to his son, Janos (1802–1860), obsessed with trying to solve the parallel postulate, another infamous math teaser:

"For God's sake, please give it up. . . . [It] may take up all your time and deprive you of your health, peace of mind, and happiness in life."

But chance, we know, favors the prepared. Wiles's fascination with the seemingly simple problem was rekindled when a friend happened to mention in casual conversation that a mutual acquaintance had just proved that there was a connection between a fairly obscure twentieth-century mathematical discovery and the teaser of his boyhood. The news awakened the Wiles within, he explained later: "I have a preference for working on things that nobody else wants to or that nobody thinks they can solve."

As can be learned in detail in several fine accounts, including Simon Singh's *Fermat's Enigma* and Amir D. Aczel's

Fermat's Last Theorem, the Princeton professor immediately threw himself into a secret life, literally hiding in his attic office to work on the puzzle so that colleagues could not guess what he was up to (and seize the laurels first) and gradually bringing together several recent, complex discoveries in order to achieve his goal. "Too many spectators would spoil the concentration," he would say when it was all over. "I discovered early on that just a mention of Fermat immediately generates too much interest."

Here we will only briefly skip through the steps that logically, though not chronologically, proved that Fermat was right. That Wiles's proof would require some two hundred pages may or may not, as those steps will indicate, also prove that the French genius was understating it (not exaggerating) when he noted that his margin was "unfortunately too small."

In 1954 two young mathematicians at the University of Tokyo, Goro Shimura and Yutaka Taniyama, met cute, in Hollywood lingo, when both needed the same issue of a German-language mathematics journal. More melodrama would follow. Taniyama, physically frail, an eccentric famous for odd dress and messy personal habits, eventually killed himself in 1958, only to be followed in suicide by his fiancée.

But after their first meeting the brilliant young pair of number theorists, both in their midtwenties, collaborated to develop a concept called the Taniyama-Shimura conjecture. They were fascinated by modular forms, which are symmetrical objects existing in four dimensions—that is, in hyperbolic space.

Modular forms, discovered only in the nineteenth century, can be found in different sizes and shapes. They cannot be drawn in three dimensions or even reliably imagined, but they can be described and analyzed mathematically. Modular forms can also be arrayed in series, depending upon the number of defining characteristics each form possesses, from 1 to infinity. This brief and cautious description of modular forms should be enough background to help us see, though vaguely, how the two colleagues in postwar Japan would bemuse the world's mathematicians.

They would suggest, or conjecture, a link between modular forms, which were thought to exist in their own exotic mathematical habitat, and a branch of mathematics known and explored profoundly since ancient times, the elliptical equations based upon the elliptic curves that grad student Wiles would study. Simply put, Taniyama and Shimura theorized, probably based upon an inspired hunch by the former, that all series of elliptic equations were mirrored by a corresponding series of modular forms. In other words, a specific modular form in the complex plane of hyperbolic space, those four unimaginable dimensions, would contain all solutions to a specific series of elliptical equations.

By that reasoning, any possible elliptic equation would be included in a modular form. If it could not be included in a modular form, it could not exist. Mathematicians were startled, puzzled, and, generally speaking, not convinced. The conjecture (also known as the Taniyama-Weil-Shimura conjecture, because of a not-unusual academic dispute over credit) was seen to be a possible clue to solving Fermat's challenge by some, but neither the conjecture nor any connection to the infamous last theorem was immediately proved.

That began to change in 1984 when a German mathematician, Gerhard Frey, rearranged Fermat's Pythagorean triple into an elliptical equation, but based upon the supposition that Fermat's last theorem was wrong. In other words, Frey's equation directly contradicted Fermat; since it was an elliptical equation, it would exist in a modular form if it was accurate. And if it existed in a modular form, as the Taniyama-Shimura conjecture theorized, it would be correct and Fermat would be wrong—*if* the Tokyo conjecture was correct. This step would become basic to awakening Wiles's interest. Frey was saying, in a roundabout way, that affirmation of Taniyama-Shimura would be affirmation of Fermat's last theorem.

A little repetition here: The Prince of Amateurs claimed that he could prove that n in the Pythagorean triple could never be a whole number greater than 2. The Japanese duo conjectured that every elliptical equation has an equivalent modular form. Frey, stating the opposite of Fermat's theorem (i.e., there could be a solution with a whole number n greater than z) as an elliptical equation, proposed that, following Taniyama-Shimura, the impossibility of Fermat's equation having a modular equivalent would show that it was inaccurate. Therefore the mathematical statement of Fermat's last theorem—no solution—would be proved correct. Frey's elliptical equation could be plotted as a so-called Frey curve.

Ken Ribet, the mutual acquaintance previously mentioned by a friend to Wiles, first thought that Frey's notion was some kind of prank when he heard about it in 1985. Moreover, unlike Wiles, he was not in the least attracted to Fermat's challenge, since he felt "nothing further of real importance could be said" about it. Then he caught the virus, or a variant of it: still dismissive of the Fermat puzzle, he was piqued by the bizarre Frey curve. In 1987 he proved that the German's elliptical equation was not modular; therefore, if the Japanese youths were right, so was Fermat. Proving their conjecture correct—that every elliptical equation is modular—would prove the last theorem correct.

You will note that quite a mound of marginalia has been heaped up here after all of these efforts, and the proofs of each of these steps required pages and pages of text and equations. Still, mathematicians had not yet found affirmation of the missing (or fantasized) "marvelous proof."

But Wiles, thanks to Ribet, had returned to the chase, concentrating on little else professionally for the following six years. He soon saw that he did not have to prove that Taniyama-Shimura was accurate for all elliptic equations—in other words, that all elliptic curves are modular—but only that a certain series of equations is always modular. Even so, he met insurmountable obstacles, despite brilliantly original use of discoveries made previously by other theorists.

By 1993 he decided to bring in a trustworthy confederate, Nick Katz, another Princeton number theorist. Katz, to use Lyndon Johnson's famous phrase, was "a man you could trust to go to the well with you." He kept mum for months about his friend's quest. So did a second colleague brought in toward the end to aid the wholesome conspiracy. But it was a moment alone, as was perhaps fitting, that finally provided Wiles the inspiration he needed.

Chancing upon an observation in a recent mathematics paper, he realized that his proof that certain elliptic curves were modular could be applied to another set as well. That rounded up the last of the equations necessary to show that the Taniyama-Shimura conjecture was correct, so far as Fermat's last theorem was concerned. Put another way, Wiles proved that the Frey curve, the opposite of Fermat's last theorem, could not exist; the Frey curve fell within the class of elliptical equations that must have a modular form.

If the Frey curve was impossible, then its opposite was true: No Pythagorean triples with n as a whole number greater than 2 could exist. Wiles, though it would eventually require those two hundred pages of text, affirmed Fermat.

"There is a certain sadness in solving the last theorem," he would write. "For many of us, his problem drew us in, and we always considered it something you dream about, but never actually do."

But, as suggested earlier, Wiles's triumph, which made worldwide headlines in June 1993, was not destined to be a one-acter. He had read his paper during a series of three daily lectures at a conference at Cambridge, his audience slowly growing from a couple of dozen to SRO. The experts had foreseen that his longwinded abstract proofs were heading in one direction. Wiles, bespectacled, balding, and 40, nonetheless pulled off a climax worthy of Beethoven: "And this proves Fermat's last theorem," he said at the close of the third talk. "I think I'll stop here."

There was, of course, as in the feel-good TV movie of the week, sustained volcanic applause after hushed silence. Unfortunately, a couple of months later, Wiles might have been moved to say *"Et tu, Brute?"* The loyal Katz, carefully going over every formula and derivation in his friend's paper word by word, symbol by symbol, found a crippling error. Soon other experts around the globe caught the same flaw and piled on. In effect, the error destroyed the proof, since it suggested that Wiles had not included all of the elliptical equations necessary to make ironclad linkage between Fermat and Taniyama-Shimura. Some equations had not been corraled in the pen.

Understandably, the second act was pure misery for Wiles. He acknowledged the problem in a dour e-mail message to other number theorists; he went back to his last. Finally, drawing upon yet another theoretician, he corrected the error. (Don't feel shortchanged here; as with much of the Wiles achievement, only specialists within the larger math community can follow the reasoning with certainty.) With not quite so much fanfare as his lectures, an article published in May 1995 closed the deal. Fermat was right. In the third act Wiles won out. Today, as seems right, he serves on the advisory board of a contest called "Millennium Prize Problems." Prizes of $1 million apiece are being offered to the first clever minds to solve seven classic mathematics stumpers.

To solve a seventeenth-century mystery, Wiles had resorted to theorems and even branches of mathematics that had not even existed back then. He himself called his work a twentieth-century proof. No one believes that Fermat could have used anything like Wiles's reasoning, especially moving from Taniyama-Shimura to Frey to Ribet and many more. We know that Fermat was right; we are unlikely ever to know the nature of his "marvelous proof," or if it existed, or if, existing, it was error-free.

The Wiles story is nonetheless a satisfying narrative. Still, it is not the defining mathematical achievement of the twentieth century. It affirms the provability of a problem in number theory; it affirms that there is an eternal truth in abstract logic.

Much earlier in the century a young mathematician, perhaps the least known authentic genius of modern times, offhandedly came up with a discovery that unnerved mathematicians, philosophers, and virtually anyone else who has encountered it. Few rational human beings want to believe that it is demonstrably true that it is impossible to attain comprehensive, contradiction-free knowledge.

Kurt Godel

Of Kurt Godel it cannot accurately be said, as the critic John Jay Chapman wrote of Ralph Waldo Emerson, "Great men are often the negation and opposite of their age. They give it the lie."

On the contrary, in 1931 the 25-year-old Godel surprised the world of his day by representing it so aptly. Consorting with the most prescient, daring thinkers in Vienna, a city already rocked by the fascist influences that would soon gain power, he proved that mathematics has unexpected gaps or blind spots. Much as Ludwig Wittgenstein was exploring the limits of the ability of language to describe reality, Godel found that the language of math had limitations, too.

Godel, an otherworldly young man and dedicated hypochondriac who also was something of a lady-killer (he horrified his bourgeois family by marrying his lifelong mate

Adele, an older woman who was not only a divorcée but also a nightclub dancer), was working to save mathematics, not shake it to the foundation. Logic led him to illogic.

In the previous century, after 2,000 years of unquestioned certainty, the hundreds of theorems of plane geometry, set down in full by Euclid in ancient Greece and taught without change wherever geometry is learned around the world, suddenly came under fire. As you recall from ninth or tenth grade, probably, two parallel lines drawn in a plane never meet. They remain the same distance from each other into infinity, given the imaginary infinite plane.

The Euclidean theorem is more specific: If a point is separate from a line but in the same plane, only one line parallel to the first line can be drawn through that point. (We're talking abstract conception here, not physics as redefined by Einstein's space-time or any other real-world context.)

But some nineteenth-century geometers got playful. It might be possible to draw two different lines through that point in the plane, each of them parallel to the original line. Or, looked at in another way, it might be that all so-called parallel lines finally meet in infinity. In other words, parallel lines do not exist.

You learned differently, even after these nineteenth-century perturbations, because Euclidean geometry is not only practical and consistent with the much-abused concept common sense, but also because it is consistent as a system within itself. The five volumes Euclid devoted to his geometric theorems carry no contradictions within. Non-Euclidean geometries can make the same claim, however; whatever their value as practical tool or guide to universal truths, they are self-consistent. To a certain kind of mathematical mind, that fact is annoying. How can something apparently nonsensical be so constructed that it stands so firmly on its own two feet?

Gazing from down the hall, mathematicians began to worry about their own bailiwick. "Logic," said a well-known German mathematician in the early twentieth century, "is the hygiene the mathematician practices to keep his ideas healthy and strong." Could there be such mayhem on the horizon in their own field as in geometry?

You have guessed the answer: Yes.

Back to the Greeks. In systematizing logic, Aristotle laid down the laws of deductive logic in the familiar form of a syllogism: If all x's are y, and z is an x, then z is a y. Here's how it works:

All men are mortal. *(first premise)*
Godel is a man. *(second premise)*
Godel is mortal. *(conclusion)*

Without going through these steps, we continually make assumptions in exactly this fashion, if we are generally reasonable; the assumptions are only as sound, of course, as the premises upon which they are based.

But deductive logic fails, it seems, in certain situations that seem innocuous enough on the verbal surface. "All Cretans are liars," said Epimenides, supposedly, in one famous ancient Greek example. Fair enough, except that Epimenides himself was a native of Crete. Was he lying, and thus telling the truth, or telling the truth, and thus lying? Don't break the furniture. There's no logical answer.

Try another. In a certain imaginary village, the barber shaves all those who do not shave themselves. Who shaves him?

Or another. Bertrand Russell, the great and droll English mathematician/philosopher, suggested writing on a piece of paper "The statement on the other side of this paper is false." Turn it over, and the other side says the same thing. "It seemed unworthy of a grown man to spend time on such trivialities," he commented in his autobiography, "but what was I to do?" What he was doing, to the consternation of others, was demonstrating that certain apparently simple contradictions are not amenable to traditional logic.

In a famous public challenge to his colleagues in the math world in 1900, the German David Hilbert highlighted twenty-three problems that, in his view, had to be solved to keep mathematics on an even keel. In short, mathematics must be proved to be comprehensive, capable of answering every possible question, and consistent, with no statements that could be proved to be both true and false. This is, after all, the question that arises from Russell's "trivialities": Can it be proved that mathematical reasoning is 100 percent and eternally reliable? No Cretan liars allowed, since that conundrum allows for two mutually contradictory possibilities.

"Every definite mathematical problem must necessarily be susceptible of an exact settlement," Hilbert explained at the time. "Either in the form of an actual answer to the question asked, or by a proof of the impossibility of its solution." Put another way, if a group of axioms were put together to produce a formal mathematical system, or set, that system must be consistent. It must make logical sense.

Though perhaps worried a bit by the monstrous geometries given birth and form over the previous decades, Hilbert certainly felt that this challenge, the second on his list of puzzles, would have a positive answer. In fact, it would probably be difficult to find many well-read laypersons today who do not assume, in some words or other, that a math problem has only one answer, no matter how strenuously they may argue with the IRS to the contrary.

But Godel proved otherwise with his "incompleteness theorem," announced in 1931 in an article, "On Formally Undecidable Propositions of Principia Mathematica and Related Systems." "On the Richter scale of mathematical discoveries," Paul Hoffman wrote in *The Man Who Loved Only Numbers* (a portrait of mathematician Paul Erdos), "Godel's was a 10."

Back to Crete: A simple version of the Epimenides conundrum, known also as the Liar's Paradox, is the sentence "This sentence is false." Within its own very small frame of reference, it embodies a contradiction. Godel came up with a slightly different reading: "This statement is not provable." Like the ancient riddle, it is contradictory: If provable, it is true, but if true, it is not provable. This Godel sentence G was his scalpel in his evisceration of axiomatic set theory.

Using a system he developed called "mirroring," Godel translated the sentence G into arithmetic language. The consequences were astonishing: In any consistent formal mathematical system that includes all rules of ordinary arithmetic, it turned out that the mathematical equivalent of his sentence G will be included. For that reason, then, the formal system includes an inconsistency; the rules of the system are unable to prove or disprove the statement.

These discoveries produced two "theorems of undecidability":

1. If axiomatic set theory is consistent, there exist theorems that can neither be proved nor disproved.
2. There is no constructive procedure that will prove axiomatic set theory to be consistent.

The math involved, as you suspect, is not easily accessible, but Godel's work can be followed in sequence. He also showed that if the arithmetic system is augmented with new ideas to make the sentence G provable, once and for all, the new material will produce its own pesky sentence G.

Finally, he translated the concept "arithmetic is consistent" into mathematical language and showed that it could not be proved. Every system will include a true statement that cannot be proved; therefore every system will be incomplctc.

"Every mathematical problem can be solved," Hilbert had asserted. "Search for the solution, you can find it by pure thought, for in mathematics there is no *ignorabimus.*" Godel *proved* him wrong. No complex system of mathematics can be proved to be consistent. Math is more complex than the language used to describe it, perhaps as human experience is much more complex and ambiguous than any of the human languages we have developed.

"God exists since mathematics is consistent," a famous number theorist said, "and the Devil exists since we cannot prove it." Hilbert had been more determined. He decreed that his tombstone read, "We must know. We will know."

For many mathematicians, Godel's discovery in abstract mathematics was as troubling, peculiar, and diverting as Heisenberg's uncertainty principle in the tiny but physically real world of subatomic particles. Did it matter very much? Probably not; after all, the man himself had not been able to isolate one of the "undecidable" statements that his theory postulated. They must be rare and irrelevant. Did 1 plus 1 not still equal 2, throughout time and space? Of course. On a day-to-day basis at the chalkboard or at the computer, Godel's discovery would seem irrelevant. And in fact, then and now, most working mathematicians decided that Godel's work, though intellectually fascinating and conceptually surprising, was of more moment to philosophy than to arithmetic. As Rudy Rucker has written in *Mind Tools,* the incompleteness theorem "became the private property of the mathematical logic establishment, and many of these academics were contemptuous of any suggestion that the theorem could have something to do with the real world."

It was not until 1963 that a young Stanford mathematician, Paul Cohen, contacted Godel with surprising news from reality: He had created a way of proving that certain specific questions, if only a small number of them, are forever "undecidable." One of them, sad to say for the eternal rest of Hilbert, who died in 1930, was on the famous list of twenty-three challenges.

Godel lived long but not happily, by most standards, and did not produce any new work as significant as his "principle of incompleteness." Escaping the Nazi annexation of Austria, he came to the Institute for Advanced Study at Princeton in 1940, but only after raving at his citizenship hearing that the U.S. Constitution could be interpreted to allow the establishment of a dictatorship. One acquaintance believed that his increasingly strange behavior indicated that he had simply lost confidence in his abilities. Though walking to his office daily in the company of Einstein, Godel was essentially a recluse, preferring to talk with a few colleagues on the telephone and paranoically having his wife, Adele, spoon-feed him when he fantasized that his food was being poisoned. This was a strange fate, it would seem, for a man who described himself as a realist and who had once tried valiantly to prove mathematically the existence of God. At age 71 Godel stopped eating entirely, it appears, when Adele was recovering from surgery in a nursing home, and he died, apparently starved to death, on January 14, 1978.

If this strikes us as a sad or a meaningless death, Godel might not have agreed. If he had found some fuzziness in the operations of math, he also believed and argued that everything in the world, everything in the universe, has a meaning, much as everything in science, he felt, must have a cause. Despite the limits he discovered to deductive reasoning, he believed that human knowledge continually expanded. If this life seems not to have a clear purpose, he reasoned, then its purpose must be to lead the way to another form of existence. Nothing discovered after his death, however, suggests that he intentionally set himself upon that road.

7

Mendel, Watson, Crick, and the Human Genome

Another Prince of Amateurs, Gregor Mendel, also leaving work that would not be read or heeded during his lifetime but without Fermat-like conundrums to cloud the sense of it, found in a monastery garden how the generations of any life form leave their legacies down the corridors of time. (As it happened, Mendel and Godel both came from Brno, the capital of Moravia at the time.)

For some, Mendel's insights into the family tree of peas, along with Watson and Crick's twentieth-century discovery of the structure of DNA, raise fundamental questions about the nature of personality, and of free will and responsible action. Each one of us is not quite as individual as a snowflake, but constructed and limited, perhaps, by complex maps made of thousands of selected particles, copied over and over again uncountable times in our somewhat mandated lives, and often subject to mistakes and errant throws of the dice.

If we are stable entities, what have we to do with that stability? To the ancient Greek playwrights, character is fate. To what extent can it be said that our genetic heritage is our fate, not only in the names of the inherited diseases that may consume us but also in the twilight fears that may literally be inherent in our personality disorders—in other words, in both body and soul? If we are our DNA, it is worth finding out more about it. Indeed, the race for more knowledge has become feverish in the early twenty-first century.

If, as we think now, previously nomadic, hunting, and gathering humans learned to settle down to agricultural life at least 10,000 to 12,000 years ago in several parts of the globe, they must have learned in due course how to breed plants and animals in ways that would optimize the traits they deemed desirable.

Something of this knowledge informs the ancient story of Jacob, Laban, and the striped, spotted, speckled, and black lambs and goats in Genesis 30. Laban, as deceitful as a publisher, agrees to give his son-in-law all of his own livestock with these markings but has his sons spirit them away, leaving only pure white animals. Jacob, as desperate to survive as a writer, uses a combination of folklore and ancient animal husbandry to remedy this injustice. Knowing that the strongest mothers are likely to produce the strongest progeny, he shows them tree branches that have been stripped to create dark-and-light patterns. Seeing these patterns (the folklore aspect), the stronger mothers give birth to strong speckled lambs and kids (traditional husbandry, presumably) that become Jacob's. The weaker mothers, who are not shown the cut branches, produce weak white lambs and kids for Laban's share. What the lords of Mesopotamia may not have known was birth knowledge out among the clans.

Perhaps coincidentally, it was someone who knew his Bible very well who first set down for other scientists what the herdsmen in the fields had known for millennia. If only someone had asked. . . .

It took Gregor Mendel, the bookish son of a farmer who bred and grafted fruit trees in their home village in northern Moravia, to provide methodical technical analysis of the kind of knowledge about breeding that had been passed down the generations of those who cultivate and herd.

His sensational breakthrough did not happen overnight. Because his father was disabled in a farming accident, Mendel joined the Augustinian religious order to finance his studies, stumbled in his first job as a parish priest, and flubbed an exam to qualify as a science teacher. Considered something of an amiable if mathematically gifted loser, the 34-year-old friar began experimenting with peas in the monastery garden in 1856. By the harvest of 1864, his meticulous analyses of the characteristics of thirty thousand pea plants and their produce was systematic enough to provide the foundation of today's genetic theory.

His revelation was the result of going one step past the obvious. Perhaps because he already knew something about their compatibility, he selected seven pairs of pea varieties to breed with each other. (From his father's orchard, he would have known that the progeny of fruit trees is not predictable; hence, the grafting.) In the following generation, as farmers surely knew, the offspring of garden peas with white seed coats and peas with gray seed coats—that is, the resulting hybrids—would all resemble only one of the parent plants. For example, all of them would have gray seed coats, as if the influence of the parent with a white seed coat had been completely expunged. The rule followed just as inexorably for hybrids produced by the other six experimental pairings. There was no intermediate breed—for example, no light gray.

What Mendel found in the next generation, however, echoes, in the anecdotal sense, comments you have heard expressed with various degrees of satisfaction, amazement, or horror in the maternity ward. Yes, looks and other characterizations do "skip a generation," only one reason why the infatuated are advised to gaze closely at a prospective mate's parents. Or to move in the other direction, who among that gleaming crowd of 18-year-olds in a faded photo from another time could possibly be Granddad? Not one of them could have desiccated into—except, yes, the one who looks and grins wickedly just like first cousin Herb.

So, too, with garden peas that could have either dwarf stems or tall stems, wrinkled seeds or perfectly round seeds. If the tall stems characterized the second generation, in the third generation they reappeared in about three-quarters of the progeny; the other one-fourth had the dwarf stems of the previously suppressed parent. In other words, the third generation would be approximately 75 percent tall-stemmed, 25 percent dwarf-stemmed—a ratio of about 3 to 1. In fact, Mendel's fastidious records of thousands of pea plants yielded a ratio of 2.98 to 1.

All of this counting unveiled a simple but essential (and, you will expect, counterintuitive) fact about heredity. It had been assumed that the offspring of two parents is a fusion or a fluid blend, like Ovaltine. Somehow, two different integrated beings seem to produce, at a glance, a being integrated in a new way. How, then, does a vanished characteristic suddenly reappear in the offspring of the offspring? Because, Mendel saw, the ingredients remain separate and distinct, even if not detectable, until they flower again in a succeeding generation. They do not blend or fuse. Each of the characteristics must be carried in an indivisible little packet. The atoms of this new atomic theory of biology, which Mendel called "factors," are now called genes. A simple organism like a virus, though capable of devastating nations, may have no more than ten genes. The fruit fly, who became the hero of genetic research throughout the twentieth century and down to today's most recent news flash in genetics research (owing, in large part, to an astonishing devotion to procreation), has ten thousand genes. Each of us humans is made from about thirty to forty thousand genes (a surprising discovery in 2001, since researchers had long believed we should have upward of a hundred thousand, and some start-up companies were claiming to potential clients and investors that they had indeed found about that many). Because genes for so many different characteristics are present in such large numbers, Mendel reasoned, it looks as if heredity is a blend. It is not.

An inherited characteristic, he surmised, is the combination of two hereditary units, each contributed by a reproductive cell from one parent. He had found that between two characteristics, one would be "dominant"—that is, the one that appeared at a ratio of 3 to 1 over the other, which was "recessive." These ratios occurred independently of the rates of inheritance of other characteristics.

Mendel was right on the money, but he and his lucid, convincing papers describing his experiments were ignored, most notably by the academy. In 1900, some sixteen years after his death, his work was rediscovered and verified virtually overnight. The biological community was at first riveted for a reason that should seem somewhat antique today—Mendel's model could be used to argue either for or against Darwin's theory of natural selection. It became clear early in the twentieth century that Mendel's work actually affirmed Darwin's. You would think that this was old news, except that six of ten U.S. respondents in the year 2000 told pollsters that "creationism" should be given equal time in public schooling along with evolution.

Whether Mendel would have been amused we cannot know, but he did not seem amused by very much, unfortunately, toward the end of his life. As abbot of his monastery from 1868, he turned from science to administration, evidently concentrating his precise intelligence on trying to get tax relief for his monastic community. Situation hopeless, of course. Hubble, Einstein, Bohr, and all of the big bang theorists arise from within select professional communities;

Mendel joins Wegener and Fermat on the outskirts. Good news for cranks? Probably not. The vindication and elaboration of the painstaking friar's revelation would come, as has almost all lasting work of science in the past 100 years and counting, in laboratories supported by ponds of money, whether from government, academia, big business, or some combination of the three. Proof these days of brilliantly simple theoretical breakthroughs requires teams of workers, time, and technologically elegant instruments.

Checking through the latest book about a hot topic for a previous chapter, I began to wonder about the state of my brain cells. A certain basic astronomical relationship seemed oddly unstable, changing its value chapter by chapter. Finally, a call to a physicist at a research university confirmed my guess: the chapters had been assigned to anonymous grad students working under the putative author. No one, apparently, had seen fit to reconcile such glitches in the text. Thomas Hunt Morgan, a Columbia University professor, was a much cannier exploiter of graduate talent in the first decade of the twentieth century, organizing legions of eager young experimenters in the search for the mechanism that passed on inherited characteristics.

From 1907 onward his team studied the inheritance patterns of the fruit fly, *Drosophila melanogaster,* which can produce two hundred to three hundred progeny in its 2-week-long life span. That adds up to twenty-four new generations per year. Four years before, the biological community learned that inherited characteristics had something to do with chromosomes. These rodlike structures, which are made from protein and nucleic acid, are found in pairs within the nucleus of every animal cell. Each animal parent contributes half of the pair in the cell of an offspring—that is, half of the information that will create a specific characteristic. Also, each parent is contributing only half of its own genetic information to the cell of the offspring.

Aside from speedy generations and its renowned zest for reproduction, *Drosophila* made a good test subject because it has only four chromosome pairs. Just as Mendel predicted, traits typically appeared as dominant or recessive down the billowing family trees, typically in a 3-to-1 ratio. Within the chromosomes, genetic information was being passed down that defined the fruit fly as unique among all species on Earth. Mendel's factors, or genes, were carried by these chromosomes, biologically informing each fruit fly how to become a fruit fly.

But something unexpected became noticeable in Morgan's orderly experiments. Every now and then, from no readily discernible cause, a fly was born with an exotic new characteristic. As Darwin had theorized, the trait disappeared in succeeding generations if it was not helpful to survival in *Drosophila*'s rather peculiar home environment, a university lab room. But when a white-eyed male fruit fly suddenly appeared among his characteristically red-eyed brethren in January 1910, it was possible to make a rather dramatic test. The progeny of the white-eyed male and a normal red-eyed female were all red-eyed, but their grandchildren were red-eyed to white-eyed in the Mendelian ratio of 3 to 1.

That explained the future, but what about the past and its ancestry? Where did the white-eyed fruit fly gene, the hard, unchanging quantum of hereditary information, come from? Morgan, who would become known as the father of American genetics, did not find the answer. As we unfortunately know from the atomic era and media sci-fi silliness, it is mutation. We take the concept for granted; a century ago, the discovery was a huge step forward in understanding inheritance and evolution.

A colleague of Morgan's, Hermann Muller, found that such mutations as white eyes in fruit flies occur because of chemical changes in genetic structure. His experiments showed that radiation with X rays could cause such changes; it was clear that they could also occur spontaneously.

Either way, a change in the information provided by the two parents could produce a characteristic in their offspring. As of flies, so is it of all other creatures brought forth, live or stillborn, in the tree of life. The genes that carried information must, since they can mutate, have a structure that can be altered.

The search for that structure was the most perfervid challenge to biological science in the first half of the twentieth century. Some would argue that the revelation of genetic structure had even broader significance than that. "It is impossible to argue with someone," wrote Peter Medawar, "so stupid as not to realize that Crick and Watson's discovery was the greatest of the century." The winners themselves, Francis Crick and James Watson, greatly aided by researcher Rosalind Franklin, have perhaps been best described by Michael Lerner: "In science and other intellectual endeavors, there are two kinds of people: the architects and the bricklayers. There are a lot of bricklayers, and they do essential work, but Watson and Crick are among the few real architects of science."

Aside from Morgan's students, the lesser contributors to quantum theory, and a mathematician or two along the way in the solution of Fermat's irksome last theorem, we have not often paused among bricklayers in this book. Einstein

Francis Crick and James Watson

didn't really need them; genetic theory did. Mountains of material were assembled page by page in labs and offices in many different research centers before the grand, overarching Watson-Crick theory was possible.

For example, the DNA molecule, or deoxyribonucleic acid, was isolated in animal cell nuclei and suspected of taking part in genetic inheritance. It existed in each of the adult human body's up to 100 trillion cells. (As we know now, this stuff is tightly coiled; if one strand of DNA could be pulled straight, like a sprung Slinky, it would stretch out to a length of 8 feet. The uncoiled length of all your DNA would be twice the distance from Earth to the Sun.) Hair cells, skin cells sloughed off by exfoliants, blood cells—whatever image works for you—there are some 200 different types of cells working together to keep you intact and alive and growing/declining. Although they vary in size, about 250 average-size human cells could be bunched together within the period that closes this sentence. Run all those stats backward—the cell's minuscule size, variety of possible forms and functions, and huge aggregation—then consider this: each of these tiny constituents of our physical being contains DNA. That discovery, widely accepted by biologists at the beginning of the twentieth century, was yet another important step. Researchers eager to find the secret of life came to believe that understanding DNA was essential to understanding Mendelian genetics, for the simple reason that we are crawling with the stuff.

Another important step was the recognition that the manufacture of proteins in each cell was somehow directed by the still-mysterious genes. Furthermore, it was learned by the chemist Linus Pauling in 1951 that some of these proteins had a helical structure—that is, they were coiled strands of material. He made this analysis of protein shape by X-raying samples.

Following Pauling's lead, Rosalind Franklin, one of science's unfairly neglected achievers, began X-raying DNA at King's College in London. Using the latest techniques in a rapidly improving technology, she was able to make the first photo of the X-ray diffraction patterns of a DNA molecule in 1952, pinpointing a fiber of DNA from a calf's thymus gland. The image suggested that the molecule might have some sort of helical structure, but much had to be inferred, perhaps as in our least successful fetal sonograms half a century later. This photo, which would became famous to biology students ever afterward as "exposure 51," looks something like a large fuzzy X made from bars spaced apart equally on each leg. Franklin could not reconcile the image with Pauling's notion of a helical structure.

Working on a rival DNA project at the Cavendish Laboratory up the rail line in Cambridge, Watson, also trying to find affirmation of the great chemist's theoretical structure, saw the Franklin photo in 1953 and felt the clouds blast away in a flash of almost violent insight: "The instant I saw the picture my mouth fell open and my pulse began to race," he would recall long after that very moment led him to worldwide fame.

What Watson saw may now seem absolutely obvious, just as in visual puzzles when, once you get it, the image flips from being two silhouetted profiles to being a Grecian urn and back and forth on and on.

Suffice to report, it was not immediately obvious to other experts who first saw it. In this case, perhaps, chance favored the prepared eye and the unusually open mind—to wit, Watson guessed that he was looking at DNA directly from the top, and the cloudy X was in fact the outline of not one helix but two, closely and regularly intertwined around each other. The structure of DNA is a double helix, he reasoned. He'd found the first part of the answer that would inspire Medawar's remark.

He and his Cavendish colleague Crick agreed on this image, sharing the news with other scientists at their regular Saturday afternoon lunch at a Cambridge pub called The Eagle. Then they quickly worked out the details, refining their conceptual model by fiddling around with sticks and wire, of the breakthrough that would earn them a Nobel Prize.

Modestly or cautiously, Watson announced the discovery to a friend in a very low key before publication in the prestigious British science magazine *Nature:* "If by chance it is right, then I suspect we will be making a slight dent into the manner in which DNA can reproduce itself." The article itself began in a similarly unprepossessing way, announcing that their model of DNA structure "has novel features which are of considerable biological interest."

Watson and Crick showed that the 8-foot-long strand of DNA is a kind of chemical language that has only four letters. The placement of those letters on the strand conveys the information of inheritance.

That's the analogy; here's the actual structure: Each strand of the double helix is a row of nucleotides, each of which is composed of the sugar deoxyribose, a phosphate, and a so-called base. The four bases (or letters, in our previous analogy) are adenine, guanine, thymine, and cytosine. Each sequence of three of these bases on the strand, whatever their order in a row, directs the creation of a specific amino acid to produce a specific protein molecule in a cell from among the several thousand different kinds of protein molecules that can be manufactured in this way. The so-called triplet code, which governs the production of the protein that is the chemical and physical basis of all life, apparently occurs in all living creatures, but the sequences vary in species and, within species, in individuals.

To carry the specific messages for constructing specific proteins, DNA relies upon a related substance, ribonucleic acid (RNA), which copies the sequence of bases in the DNA strands and relays the information to a part of the cell called the ribosome. There, triplet by triplet, unrolling like ticker tape, the messenger RNA gives the ribosome instructions for producing transfer RNA, which in turn produces the specific amino acids (chosen from a group of twenty-two in varying amounts) that combine in a chain to form a specific protein molecule. At the end of this intricate procedure—DNA to messenger RNA to ribosome to transfer RNA to amino acids to protein—the protein chain has been constructed in three dimensions, possibly containing hundreds of amino acids, and it is given instructions for whatever task it is designed to perform. Proteins include hormones, which regulate the metabolism; enzymes, which spark functions within the cell; antibodies, which are essential to immune functions; and other essential biological molecules.

As the two strands of the DNA helix wind around each other, the nucleotides of one are bonded with the nucleotides of the other in complementary, mirroring strands. By repeating these bonds, the strands can replicate, or copy themselves, according to very strict rules: adenine and thymine bond only with each other; the same for cytosine and guanine. These connections are called complementary base pairs; that is to say, adenine-guanine-thymine-cytosine in sequence on one strand would be mirrored on the adjacent strand as thymine-cytosine-adenine-guanine, and so on in the looking glass for the entire row of three billion pairs in a strand of DNA.

To start the replication, enzymes in the cell rush over to the helix and encourage the complementary base pairs to separate from each other, unwinding the two strands of nucleotides. Other enzymes rush in (as you might expect, there are hundreds of different types toiling away in your body as you read), pushing newly produced nucleotides into mirroring positions, following the pattern of the single strand, until a new double helix is formed. After replication is completed on each single strand, there are two double helices that are exactly like the original one; each is composed of a strand from the original along with a complementary new strand of nucleotides.

Some part of this process is going on right now in every one of your hundred trillion cells, for they are constantly growing and dividing into daughter cells, continually renewing until halted by injury or disease, death, and decay. We see in the mirror or by marks on the wall or by belt sizes that we are aging; what we cannot see or feel is the constant ferment of life at the cellular level. It is at that level, too, that essential decisions are made about the type of cell. DNA is the same in every cell, whether in the blood or in the bone or in sputum—the basis for the genotyping that has rescued a startling number of inmates from death row or lesser mistaken penalties. Why, then, do the body's cells differ? The answer is simple and, if you like, wondrous: Only the genetic material that makes blood is activated in a blood cell; the rest is silent. To return to the language analogy, only the combination of four letters that forms the DNA equivalent of the word "blood" is understood. All other possible four-letter combinations are drivel.

Human cells contain twenty-three pairs of chromosomes, or a total of forty-six packets of genetic information; the DNA in each chromosome includes some thirty to forty thousand genes. Mutations, as noted, can occur spontaneously or as a result of environmental assault, including radiation, but malfunctions in cell division, which involves astronomical numbers of genes over a lifetime, are not statistically unlikely. It is astonishing that almost all cell division occurs without mishap or is repaired by the body.

Even so, some four thousand gene-linked diseases are known, quirky to deadly malfunctions that are passed down—usually unwittingly, always implacably—from parent to child. Nor are they necessarily as simple to graph as the either/or characteristics of Mendel's garden peas; Alzheimer's disease, for example, may be linked to more than one gene in one of its forms, while other forms may arise from a single gene carrying the terrible instruction to destroy personality. With this disease and many others, including arthritis and multiple sclerosis and cardiovascular disease and cancers, the parent or grandparent provides a devastating legacy, and human will has yet little to do with the outcome. In some statistically few instances, no great love and no sure care can overcome genetic programming that causes pain or madness or premature death.

This is peculiar information, at least in contrast with our ancient ideas of character and physique. In the *Iliad,* when grieving Priam convinces wrathful Achilles to release Prince Hector's body, the old king pauses for a moment, admiring the fierce Greek champion, as handsome to his rheumy eyes as a god. It is the poet Homer's point, as usual, that youth and beauty cannot last in mortal life; age and death predictably cut them down in the pattern of life.

Our view now has to be more complex, and perhaps more troubling, even if our fates are ultimately the same: some of those trillions of cells incessantly dividing, making uncountable manufacturing decisions according to instructions set up in a kind of game of chance, may even now be unpredictably going off-message, spreading out of control as cancer, or producing protein molecules that, directly on-message, will somehow bring on early-onset Alzheimer's disease, so that proud but unlucky Achilles will not only forget why he slaughtered Hector but even what a Hector might be.

As to the most physically obvious and arguable genetic distinction of all, the sex cells of eggs and sperms are exceptions to typical cell structure. Each contains only twenty-three chromosomes; when they merge to form cells that will produce offspring, the resulting forty-six chromosomes represent each parent cell equally. These paired chromosomes are usually shaped like an X; if a Y-shaped chromosome is contributed by a parent, it is paired with a larger X-shaped chromosome. The introduction of this pair causes the offspring to be male.

Such a global distinction, though elemental, may pale in contrast with the billions of genetic choices made within the other cells. In order to learn how each individual differs from all others, in order to seek out and perhaps repair those bits of "misinformation" that instruct cells to produce disease, the much-publicized Human Genome Project mapped the genome, the step-by-step DNA sequence of three billion pairs (or, back to our alphabet analogy, six billion letters) in each human cell. Surprisingly, it has already become clear that only some 3 to 5 percent of the genetic structure plays an active role in passing on genetic information. Were the inactive genes, known as junk genes, required hundreds of thousands or even millions of years ago by our hominid ancestors? We don't yet know. Since all of life is genetically programmed in much the same way, however, it seems likely that all organisms alive today and all dinosaurs and other species that have passed out of existence share and have shared a primordial helix that began replicating 3 billion to 4 billion years ago, the lead sentence in the text that would become the lengthy, mysterious human genome.

That genome, by the way, shows that there is no such thing as "race" in the popular sense of skin color or "ethnicity." There is, genetically speaking, only the human race, with many more variations within a socially determined so-called racial group than between that group and another. Any human being can physically mate with any other human being of the opposite sex. The Mendelian rules will apply to their progeny as to any other human progeny. When the Martians land, ask them if they notice race. Perhaps, since they could not build spaceships without learning to make distinctions and classifications, they would immediately say "Of course. We see that you have different races according to their heights." Or whatever.

And if you happen to feel infinitely superior when you stare into the moist brown eyes of a chimpanzee at the zoo, certain that our closest relative species is not all that close, consider this: The actively working genes in the chimp's body and in yours are 99.6 percent identical. That tiny fraction of genetic material distinguishes humans from the other apes. What is produced by that 0.4 percent? The ability to speak, to reason, to contrive a belief in individual identity and the soul?

"We used to think our fate was in our stars," Watson has said. "Now we know that our fate, in large measure, is in our genes."

Exactly how large is that measure? Are we dangerously overweight because of a genetic legacy, or because we overindulge in potato chips and slouch for hours in front of the TV set? Will genetic heritage become the ultimate "Twinkie defense," when the lawyers see that strong arguments can be made for the genetic legacy of personality disorders as well as physical genetic disorders such as cancer? Will researchers eventually be able to isolate the gene that causes its carriers to keep the plastic slipcovers on their furniture? Or is this behavior a product of environment . . . or of interaction between genetic legacy and experience?

These are serious questions that will not soon be answered, and there may be many kinds of mischief on the horizon before the answers become clearer.

At the moment it seems clear that all attempts to link a specific gene to a specific type of behavior—in other words, experiments in human behavioral genetics—have failed, no matter what you've read in the tabloids or heard on the nightly news. In those venues, studies have been promulgated that supposedly found genes that cause alcoholism, manic-depression, or schizophrenia. These conclusions have been roundly refuted.

There are good reasons to expect that such connections can never be made. For one thing, genetic determination of the structure and other physical properties of the human brain, which can be shown to be inherited characteristics, does not necessarily imply biological determinism in regard to behavior. While it is possible theoretically that genes could somehow determine or control behavior, the theory has not stood up in experiments.

For another thing, a family history of a particular behavior is not only a Mendelian genetic sequence, it is also a history of interaction between individuals and their relatives, individuals and their environment. Put crudely, researchers have not shown that an alcoholism gene was necessarily required to produce an alcoholic in a family of drunks. And even though deterministic Mendelian rules seem to apply harshly with certain rare single-gene diseases, such as sickle-cell anemia, the conditions are not manifested in the same ways in each heir and may be susceptible to environmental intervention. And for most inherited diseases, many different genes and many different environmental factors are involved; in sum, there is no such thing, despite what a television medical expert might seem to be saying in a 90-second clip, as a gene for breast cancer or for diabetes. In the latter case, as many as fifteen genes, and perhaps more, can be involved in producing the disease. Similarly, human behaviors, which might have a genetic component, are directed by our incredibly complex brain matter in interaction with the environmental factors of a lifetime. As Matt Ridley phrased it in *Genome: The Autobiography of a Species in 23 Chapters,* "The more we delve into the genome, the less fatalistic it will seem. Gray indeterminacy, variable causality, and vague predisposition are the hallmarks of the system."

Put another way, it has so far proved impossible to distinguish the effects of nature from those of nurture, a controversy that, in different forms, has swung back and forth in psychological studies in Europe and America since at least the nineteenth century. A pernicious manifestation was the eugenics movement in this country between the world wars, when blond, blue-eyed families were awarded prizes at county fairs in the heartland for their ideal breeding. During this nonsense the American Breeders Association recommended that "defective classes be eliminated from the human stock through sterilization." This was Mendelian genetics gone mad, and the Nazi excesses made it distinctly unfashionable after World War II, but it has not disappeared from private discourse and crackpot pamphleteering. It has surfaced most recently in a renewed controversy over IQ, with one side arguing that relatively lower intelligence can be attributed to "racial" heritage and the other arguing that environmental factors explain apparent differences in intellect between members of different "races."

In this IQ controversy, as in the previously mentioned refutations of gene-linked behaviors, no specific gene or group of genes has been found to affirm the conviction—for it is that, not speculation—that intelligence is inherited in the same systematic way as hair color or the ability to curl one's tongue.

Drosophila, along with other of the lesser creatures, has been enlisted in the attempt to find genetic links to behavior. So far, no stunning announcement. Because dog breeds seem to exhibit such inherited characteristics as the urge to fetch or swim or herd, a Dog Genome Project has been set up to explore the possibility of canine behavioral genetics. In time it may well turn out that certain kinds of genetic predisposition to behaviors dysfunctional or beneficent will be discovered. Why always the negative? If there is an ax murderer's gene, why not a saint's gene?

If such predispositions are found, the response may verge on the hysterical in some quarters, and it will be critically important to remember that such discoveries would not prove the case for genetic determinism. The environment of the family, and indeed the human community in all its forms, interacts with the genetic blueprint to form the human personality. In theory, genetic diseases will someday be prevented or alleviated by the use of gene therapies. In a similar fashion, should it seem likely that someone is more predisposed than the average person to violent or self-destructive behavior, the potential problem can be addressed with either genetic or environmental remedies. As molecular biologist Tim Tully has predicted, "Molecular studies of behavior will ultimately show that environmental intervention will work."

At the same time, as we all know or fear, the potential for abuse of individual rights might be tremendous if certain genetic insights fall into the hands of government and business interests; perhaps a wider understanding of the limitations of the genetic blueprint in influencing, much less determining, human behavior will help avoid such abuse. On a lighter and sillier note, the strangely abiding controversy over the potential for cloning humans is predicated upon the misguided belief that genetic determinism is omnipotent. Any such clones would develop into very different personalities because of the complexity and unpredictability of the myriad of environmental factors of a human lifetime.

8

Hominids, Humans, and the Search for Origins

What geneticists will discover in the near future can be only partially surmised, but they have already made discoveries about our human origins that, in just the past couple of decades, have completely revolutionized what experts think they can know about our beginnings as a race.

The human genome that is now being analyzed down to its last subunit appeared on Earth for the first time, perhaps, from 100,000 to 200,000 years ago on the savannas or in the forests of an East Africa very different from the hot, dry canyons and soupy swamps there now. (The first mammals preceded us into existence by about 220 million years.) Geneticists have gone back through the genetic record to discover that all humans alive today are descended from not one "African Eve," as the headlines have claimed, but from a very few primordial mothers. They and their kinsmen were not an earlier form of us; they *were* us.

When the perp gets nailed by DNA on *Law and Order,* say, the evidence comes from that unique genetic fingerprint created within the nucleus of the cell. Another type of self-replicating deoxyribonucleic acid, known as mitochondrial DNA, also is produced within the cell but outside the nucleus. It is passed down only from the mother to the offspring, then passed down in that generation by the daughters to their own offspring; in other words, it is a chemical family tree that can be traced backward only in the maternal line.

It was mitochondrial DNA that led to the recent identification of the skeletal remains in Siberia as the assassinated Romanov family, without, as it happened, any missing Anastasia or hemophiliac czarevitch; everyone in that doomed family was present and is now accounted for. Since the divine right of kings made for a very exclusive club in Europe in the nineteenth century, Czar Nicholas II and his family inherited mitochondrial DNA that could also be found in the bodies of royal cousins from England to Greece. All harked back to a very small pool of royal mothers. Researchers found matches in several places, including the veins of Britain's Prince Philip.

Some molecular geneticists today feel that the so-called African Eves, whose mitochondrial DNA lies within every human being living today, appeared on Earth 100,000 to 170,000 years ago, with 125,000 as the most generally accepted estimate. (Alternatively, at that point they began to migrate out of Africa.) As with the age of the universe, the imprecision is less important than the fact, agreed to be demonstrable, that our beginnings have turned out to be much more ancient than anyone thought possible before the last decades of the twentieth century.

These people, mind, were indeed *Homo sapiens.* If they were all much more slender than the average American (not a great challenge), brown-skinned, and dark-haired, as some molecular geneticists have argued, they were genetically indistinguishable from you, except for whatever variations within a species may distinguish you slightly from your relatives and neighbors. That is to say, you wouldn't particularly notice one of them as unusual at the local mall, but with DNA evidence, you could reliably convict him of carjacking.

But wait, you say—"appeared on Earth"?

Evidently that is exactly how a new species gets going on the evolutionary scale. It is not that one species slowly evolves over millennia into its successor—the misconception that gave rise to the wrongheaded notion of a "missing link" between *Homo sapiens* and some differently fashioned ape "ancestor." Instead, speciation occurs in a kind of leap, according to the theory of punctuated equilibrium proposed by Stephen Jay Gould and Niles Eldredge in 1972. As in the quantum world, however, the word "leap" is perhaps misleading. An evolutionary change documented in the lineage of an African snail, for example, took 5,000 to 50,000 years to be complete.

How this happens we do not yet know. Why it does, if causation is even a relevant consideration, we have not guessed. But there are markers in the mist. If recent molecular biological studies are accurate, humanlike creatures, or hominids (members of the family Hominidae), abruptly broke off from the ape family tree 5.5 million to 6.5 million years ago. On average, according to genetic and other evidence, a species endures for about 10 million years. (If that rule holds, the Sun's predictable demise is *really* not going to be a problem.)

Gorillas and orangutans come from the same family tree as we but branched off millions of years before we did. Chimpanzees, our closest cousins, and bonobos, another type of ape, became distinct species after we did; in other words, the three of us share a common ancestor whose characteristics can be guessed at by analyzing ours.

For example, humans, chimps, and bonobos all can recognize themselves in a mirror; our common ancestor must have had that capacity, a primordial Narcissus beside a still forest pool. On the other hand, we do not typically live in trees, but since both chimps and bonobos do, our shared ancestor probably did as well. We lost a trait that was passed along to the other two branches of the family, just as we became virtually hairless in comparison with these cousins. A hot candidate for our common ancestor is a tiny, large-eyed, tree-dwelling extinct creature, a kind of lemur, but the evidence is disputed.

We might consider again the end point fallacy. Are we *Homo sapiens* the completed product, the apex of the human line of development? There is no reason to think so. The world still turns; the Sun still races around the Milky Way. If we can't yet say how and why the many hominids in the millions of years before us were succeeded by the abrupt punctuation that produced us, we cannot begin to guess sensibly what factor would produce the next punctuation mark, a new race of hominids.

If we're a way point, not an end point, we benefit from the way points that preceded us—and may pass along benefits to the next team. Yet a certain kind of human being resists, or even resents, the notion that we have ancestors who were so different from us. Of course, there are the so-called Creationists who have believed that we are created in the image of a Semitic sky-god, male before female. You will still find college graduates who "argue," "My grandfather was no ape." Such people are determined to waste their minds, a provably aberrant activity in the extraordinary development of the human race.

Weird, stubborn, false pride took a different form in the nineteenth and early twentieth centuries, when anthropologists generally believed that *Homo sapiens,* defined for them as the species that created Greek democracy and designed the Chartres cathedral and produced polyphony, could not have arisen, or been created, south and east of Western Europe.

Only slowly did evidence from Africa and Asia provide a very different portrait of human origins.

At first, discovery favored the Eurocentrists. In 1856 the skeletal remains of very ancient humanlike-yet-nonhuman creatures were found in the Neander Valley in Germany. Up through 1925, so many "Neanderthals" were dug up in Western Europe—each adult skull capable of holding a human-size 3-pound brain—that Eurocentric evolution seemed a certainty. They had existed, it appeared, up until only 30,000 years ago. In France the unearthing of a slightly different hominid, dubbed Cro-Magnon man, seemed to cinch the case.

On the other hand, there was Java man, whose 800,000-year-old remains were found in Trinil, Java. This ancient upright ape, unlike any living creature known at the time, became the candidate of some theorists for progenitor of the increasingly populous Neanderthals. Both types of skull had prominent brow ridges and were oval-shaped. If Java begat Neanderthal, did the latter become the forefathers of *Homo sapiens?* And there was Peking man, unearthed near the capital of China, to be followed by other Peking men. They lived 500,000 years before.

For many, these discoveries did not provide a pleasant prospect. An Asian origin for our primordial ancestry was a blow to Eurocentrism. Moreover, the so-called Java man, a potential ape forebear rather than a hominid close to our own image, ran afoul of the "Adam complex," the conviction or hope that our earliest ancestors must from the very beginning have been more intelligent and, by our lights, more physically attractive than any of our ape cousins.

Worse news for this mind-set was to come from Africa. Let us not forget that the dominant white culture in many places had long been obsessed with the "otherness" of Africans of their own time. Even Abraham Lincoln's ambassador to France, John Bigelow, recalling a lunch with the writer Alexandre Dumas *père,* grandson of a black West Indian mother and a white French noble, remarked afterward that "the peculiar character of the African" was evident in a lack "of the reflective and logical faculties." Worse bias was demonstrated in much worse ways up and down the social levels for generations, of course, but if a diplomat supposedly waging an antislavery crusade could betray such notions, it was also true that many a supposedly objective scientist, whether measuring the size of an African's brain case or trying to demonstrate that black athletes inherit superior stamina, was convinced in the nineteenth and early twentieth centuries that a Napoleon or a Michelangelo could not have descended from African forebears.

Sometime in 1924 the truth began to emerge during mining operations in a rock quarry in Taung, South Africa. Embedded in a piece of rubble that was tossed about after an exploratory blast was part of a skull that exhibited both apelike and humanlike characteristics.

This Taung skull, at first analyzed by Raymond Arthur Dart, an academic brain researcher with no previous interest in paleontology, evidently belonged to a creature perhaps only 4 years old. Like humans as opposed to apes, it had a small canine tooth rather than a fanglike canine for interspecies fighting. Humans, from flung pebble to intercontinental ballistics missile, have used tools for this purpose. The structure of the skull indicated that the primordial toddler had walked on its hind legs, because of the placement of the opening for the spinal column at the base of the brain cavity.

Dart, vaulting Wegener-like outside his field, immediately speculated that the apelike hominid was the "missing link" between ape and humans as we know them. The skull was only large enough to contain an ape-size brain, but its jawbones seemed humanlike.

Dart was wrong about the linkage, of course, but he correctly saw that he was analyzing the remains of the first primate fossil uncovered in Africa south of the equator. Because he also knew that the skull was found in a stratum of rock at least 2 million years old, he recognized that the creature, whatever it was, was the most ancient of all human ancestors found so far. (Remember, current theory holds that our line deviated from the ape family tree only 3 million years earlier, or 5 million years ago; this was a huge leap backward into prehistory.)

With the predictable scorn for the nonspecialist, European experts rejected Dart's speculations and conclusions when they were published in 1925, but the skull was in fact the first fossil discovery of a species now known as *Australopithecus africanus* (*australis* is "south" in ancient Latin, *pithecus* "ape" in ancient Greek). Skeptics argued that perhaps because the skull was so young, it had not yet developed the essential apelike nature of its true species. Besides, no one believed that humans or their close ancestors had existed for longer than about 500,000 years. But Dart paid the doubters no mind, it seems; call it obsession or inspiration, he devoted the next quarter century of his life to finding the lone toddler's cousins.

Eleven years after his published announcement, one of his colleagues, Robert Broom, finally found an adult skull some miles from the Taung quarry. Over the next decade, skeletons of *Australopithecus africanus* complete enough to prove that Dart was correct in assuming that the species walked upright were found within a cave at nearby Sterkfontein. Broom's work and convincing analyses persuaded the academics, at last, that *A. africanus* was indisputably human, probably thriving between 1 million and 2 million years ago.

By the end of the twentieth century, more than six hundred examples of this ancient creature had been discovered. To judge from the siting of the digs, it lived in communities in close proximity to each other. These finds were enough, in the minds of many, to vindicate the prediction by Charles Darwin in 1871 that Africa was the cradle of humankind.

Such tantalizing evidence inspired the colorful, often controversial, and obviously talented Leakey family: Louis S. B. and Mary; their son Richard; and his wife, Meave. Their evident gifts at self-promotion—*père* was occasionally dismissed early in his career as the Abominable Showman—should in no way detract from their extraordinary achievements. About 6 years after Dart first handled the Taung skull, the elder Leakeys, then in their twenties, began digging for ancient stone tools in the now-famous Olduvai Gorge, a 30-mile-long stretch of the unpleasantly hot and dry Great Rift Valley in Tanzania.

The culmination of their early work was Mary's discovery in 1959 of an australopithecine, dubbed *Australopithecus*

Charles Darwin

boisei, which was virtually a contemporary of *A. africanus* but had much larger teeth and jaws. Evidently, since these hominids had no canine teeth, they chewed tough vegetation from side to side, aided by strong jaw muscles. The skull's site suggested that it lived 1,790,000 years before. In other words, more than one hominid species might have been living at the same time on the African tectonic plate as it slowly ground southward millions of years ago. More importantly, this specimen was found beside a tidy collection of simple tools; this skill is considered a distinguishing characteristic of the developing human line.

All in all, the fossil record seems to indicate that enlarged skulls and a diversity of separate species within *Hominidae* appeared more or less in sync about 2 million years ago. Molecular geneticists believe that the larger skulls show that genetic changes at about this time produced larger brains with greater reasoning capacity, including the ability to communicate with others of the same species using symbols. (Actual speech, it is thought, probably did not originate until *H. sapiens* arrived on the scene about 125,000 years ago.) Every 100,000 years thereafter, according to one speculative researcher, the hominid brain would grow by about 150 million brain cells.

How did these various species react with each other, if in fact they were contemporaries? Why did they all disappear? Did they, as we cynically expect after our own antics over the centuries, succeed in killing each other off? Did uniquely carnivorous *H. sapiens* find his cousins nutritious? Do the discoveries so far represent an incomplete or misleading record, since remnants are found capriciously because of climatic or geologic factors that favored the survival and discovery of some remains rather than others that might be more typical? These types of questions have not yet been answered fully.

The patchwork of discoveries continued in Africa, however. The oddly charismatic female australopithecine skeleton known worldwide as Lucy (and supposedly named by the researchers who found her for the Beatles song "Lucy in the Sky with Diamonds"), unearthed by Donald Johanson in a very remote desert region of Ethiopia in 1974, probably lived 3.2 million years ago. Some 3.5 feet in height, she was the first known example of yet another species, *Australopithecus afarensis.* Johanson and his team became convinced that the structure of her pelvis, combined with the shapes of her femur and tibia, proved that she walked upright.

In 1978 the indefatigable Mary Leakey discovered footprints made by a pair of these creatures some 20 miles from the Olduvai Gorge near the Sadiman volcano. About 3.6 million years ago, according to the 200-foot-long fossil trackway, two *A. afarensis* adults with well-developed arches walked side by side. These primordial footprints are, as you can imagine, the oldest known footprints to record bipedal walking, itself a currently hot topic of debate. Did we rise on two legs in order to espy our predators, gather food from branches, run more efficiently, or simply cool ourselves by exposing less of the body to the sun?

Nearby, although not necessarily made on the same occasion, are prints left by a three-toed horse, also long extinct. One may like to imagine a domestic couple walking homeward accompanied by an animal companion, hominids so very much like humans in fundamental ways far upward on the branch of the family tree, but there's no proof. In fact, scientists looking at the same set of hominid footprints have seen two adults (one following in the footsteps of the other) and a child, or a male and two females. Virtually everyone you know thinks that *American Gothic* is a painting of a husband and a wife, when it is actually father and daughter. The imagined image could show loyalty and endurance; the true image could suggest exploitation or a failed life or worse. For much the same reason, possibly, we may want those footprints to indicate stability of humanlike connection, even though we know that the species, evidently successful for millions of years, vanished.

Two years before Lucy's apotheosis, Richard Leakey found a fossil skull north of Olduvai in Kenya that bore a closer resemblance to *Homo sapiens* than did any of the australopithecines. Perhaps 1.9 million years old, its nearly complete skull indicated that the species had a brain significantly larger than those previously found. *Homo habilis,* as the younger Leakey named it, was even more adept at making and using tools, evidently, than the australopithecine found by his mother. It was also roughly the same age as her find, but some three times the age of any human skeleton found.

H. habilis was in fact close kin to a species that Richard's team discovered in Kenya in 1984, *Homo erectus,* whose skull resembled the centuries-old finds of Neanderthal skulls. Since then, other experts have classified the remains as *Homo orgaster,* a separate species. (It was not until 1987 that analysis of Neanderthal DNA proved that they are not in our ancestral line. A decade later a molecular geneticist determined that they probably split from our branch of hominids about 600,000 years ago.) The particular skeleton found was some 1.6 million years old—younger than Lucy, perhaps contemporary with *A. africanus* and *H. habilis,* older than Neanderthals. Yet such comparisons are not enough to tell us definitively about the specifics of human lineage.

For a brief time there was unprofitable debate about the Lucy family's exact role in human development. Johanson tried to make the case that *A. afarensis* was the ancestor of *H. habilis,* who was in turn the ancestor of us. From *A. afarensis,* too, arose the robust but doomed *A. boisei* and a cousin, *Astralopithecus robustus,* but that side of the family died out forever about 1 million or so years ago. (These related extinct australopithecines are called "robust" because of the heavy jaws seen in the Leakey find; in life, they were smaller than our much closer cousins, the chimpanzees.) The addition of *Homo erectus,* a possible descendant of *H. habilis* and possible close ancestor of *H. sapiens,* has filled out the story for some. *H. erectus* remains have been uncovered in 1 million-year-old sites from Africa up through parts of Eurasia but may have first appeared 2.1 million years ago.

To those who objected to this pleasant diagram, most notably the Leakey family, there were some unfortunate questions, such as the 1,000-mile-long trek required between

Hadar and the Olduvai and the numerous gaps in the fossil record. While the Leakeys agreed that *Homo* and *Australopithecus* surely have a common progenitor, they argued that this unknown hominid must have lived some 7 million to 8 million years ago, appreciably before Lucy's clan thrived in Ethiopia. (And what if, as a pair of anthropologists later suggested after analyzing her pelvic area, the comparatively leggy Lucy was actually a man? "He" would be an adult male of some other species rather than the smaller female representative of *A. afarensis.*)

Keeping the family business going after Richard lost both legs in a civil aviation accident, Meave discovered a new australopithecine in a previously unplumbed site near Kenya's Lake Turkana in 1994. This distinct hominid species, according to her team, dated back 3.9 million to 4.4 million years before. At about the same time she named her find *A. anamensis,* however, diggers in Ethiopia uncovered the remains of still another new hominid subspecies, the ramidus species, who also lived perhaps 4.4 million years ago. In March 2001, Ms. Leakey wrote that a skull she had found recently in Kenya was more likely than Lucy's family to be our direct ancestor. She named the 3.5-million-year-old fossil *Kenyanthropus platyops,* considering it a previously unknown subspecies, although some experts believe it to be a variant of *A. afarensis.* Further complicating our possible history was the discovery in Ethiopia, announced in July 2001, of partial skeletons of five specimens of a 5.8-million-year-old creature who might possibly be the ancestor we share with chimpanzees. Representatives of the ramidus species, they walked upright and lived in the cool, wet, primeval forests of East Africa. If indeed they are proved to be our ancestors, current human evolutionary theory will have to be revised. Traditionally it was thought that hominids evolved after the forests disappeared, giving way to the more congenial environment of ancient savannas.

Granted the likelihood of new discoveries and competing analyses, the current generally accepted but tentative theory of human evolution can be cautiously sketched. Some 3 million to 5 million years ago, Lucy's family, *A. afarensis,* were the only hominids in existence, and also the only mammals to walk upright on two feet. Although small-brained, these hominids probably took advantage of their unique physiques to carry their young, food, and tools from one campsite to another. They possibly learned the essential human concept of "division of labor": while the foragers roamed across the savannas to find plants to eat, another part of a community may have tended the young and protected the homesite. They probably knew how to make fires.

When enlarged skulls appeared some 2 million years ago, so did three separate species: *A. boisei, A. robustus,* and *H. habilis.* As the two "robust" species died out, *H. habilis,* with a brain half the size of yours but a very similar physique, learned to eat the meat that provided protein for a growing brain. It evolved slowly over the next 500,000 years into *H. erectus.* This hominid, with a brain about 50 percent larger than that of its immediate forebear when it became a distinct species, continued to evolve as it moved northward to Europe and eastward to Asia, where it is represented by the 500,000-year-old fossils known as Peking men. By that point its brain had grown nearly to the size of ours. *H. sapiens* would soon appear.

What the dry, dusty dig sites misrepresent is our ancient history as hominids in Africa, where perhaps the ancestor we share with chimps and bonobos lived in the branches of the forest. At about the time when we broke off the family tree, the environment may have changed to broad savannas, perhaps making it more sensible for us to walk upright.

By the time of the Laetoli footprints found by Mary Leakey, however, our hominid ancestors or cousins in East Africa found themselves having to deal with a much drier climate. As three different tectonic plates met at the Rift Valley, they pushed up cliffs hundreds of feet high that cut off the moist winds rolling across the continent from the west.

The so-called robust australopithecines, like Mary Leakey's *A. boisei,* foraged for plants and then died out. Meanwhile, the larger *Homo* was on the scene—thick of skull, skilled with stone weapons, adapted to catching and killing and eating meat, but sluggish of development. These characteristics combined to the great advantage of our hominid ancestors: a lengthy preparation for adulthood allowed this uniquely large (and growing) brain to develop its potential, the enlarged and strong skull let the brain grow while protecting it, and protein from fresh kills fueled the brain's increasingly complex activities.

Yet these advantages proved to be, in the end of each species of *Homo,* insufficient. We may have appeared in something of an instant, perhaps while some of our immediate predecessors still hunted and reasoned and developed their insular communities for survival, and we did survive, but just barely. According to one interpretation of the genetic record, we first appeared as a very small band, perhaps only a few thousand or fewer. Some researchers take that same evidence further, inferring that we may have increased in population to some 100,000, then unaccountably dwindled to only 10,000 about 100,000 years ago. Some crisis in East Africa—no one knows what—may have almost wiped us out as surely as all previous hominids were extinguished.

Why not us?

The usual answer, of course, as delivered by the winner, is our singular intelligence combined with remarkable adaptability. Millennia before former U.S. attorney general John Mitchell advised "when the going gets tough, the tough get going," we apparently did just that. Perhaps it was only 50,000 years ago that we began to spread around the world, if you believe one group of anthropologists, initiating the process that has taken a species of hunter-gatherers in a relatively mild climate to virtually every part of the globe, finally pushing aside or dominating every other species in existence.

In so doing, we learned to farm at altitudes as high as 14,000 feet; build protection for the winter in alps; fish the deep waters of the western Pacific; domesticate and breed animals for our convenience; refine our languages and thought so that communication ensured survival, as if we always guessed that knowledge is power; and continue our characteristic toolmaking, almost always making the sword before the plowshare, the missile before the space probe.

As we ventured out, we were changed. In sunnier climes our skin darkened to protect us, as it lightened in northerly areas to soak up the sun's health-sustaining vitamins. We grew chubby where it is usually cold, since a sphere conserves heat better than any other geometric solid, and slender where it is torrid, in order to promote faster perspiration.

After our dispersal into most corners of the globe, we lost contact with each other. Few took round trips as our migrations speared in multiple directions out of Africa. (Even as recently as the Irish folk song "Danny Boy," a trip to America to find work was a kind of death for those left behind, causing the song to be misinterpreted widely as a lament for casualties of war.) The Norse did not exchange genes with the Mameluks. It was the substantial diversity of the human genome that allowed for traits to gain a hold by natural selection and for random changes to appear.

Thus we probably became the different physical types that have been called races. As Edward O. Wilson notes in *The Diversity of Life,* "Anthropologists, like biologists, have now largely abandoned the formal subspecies concept. They prefer a convenient shorthand to designate a certain part of a population with reference to one or two traits." Such traits are studied today in relation to the geographic areas in which they are found and their possible optimization, in such contexts, of human reproduction and survival.

At the same time, this portrait of human development is not without controversy, both scientific and sociopolitical. Some researchers argue that various populations of *Homo sapiens* arose in several parts of the world at more or less the same time. In that case, the main three population groups—Africans, Asians, and Europeans—became what they are because of the differing circumstances of their development. This idea has not gained wide acceptance among professionals. Meanwhile, proponents of the "out of Africa" thesis hoped to chart the patterns of migration in the proposed Human Genome Diversity Project, which would analyze genomes collected throughout the other continents. This project never got off the ground because of objections that Third World, impoverished peoples to be targeted would benefit more from improved healthcare than from participation in such an experiment.

A similar ethical issue is currently being raised in the mapping of the human genome. On the one hand, it is possible that the major population groups, however they arose, have differing rates of genetic predisposition to such diseases as diabetes, cancer, and schizophrenia. That knowledge could be useful in allocating resources effectively. On the other hand, such information, it has been argued, could stigmatize population groups, feeding into the traditional simplistic ideas of ethnicity that scientists have tried to undermine with their research. Historically, we have not been very wise as the human race in dealing with such information as, say, an Asian group might be slightly more likely to have a genetic predisposition to a specific disease than Caucasians.

By about 40,000 years ago, it is now thought by some experts, we made our way into Europe, arriving in time to witness, at least in some cases, the decline and the death of Neanderthals. (The Cro-Magnons found in France were among the vanguard of *H. sapiens* pushing westward toward the Atlantic.)

Earlier we had lived alongside the mysterious Neanderthals, it seems, in the Mideast, where a rude attempt to bury a dead Neanderthal infant under a rocky cliff has been interpreted to suggest that this species may have had a notion of an afterlife. In Central Europe we were still neighbors until about 30,000 years ago, when the Neanderthals disappear, perhaps because of an inability to adapt to severe climatic changes or possibly because of a tendency to hunker down rather than migrate in an environmental crisis. Recent news accounts of fossils that might have been the progeny of interbreeding between the Neanderthals and us—two distinct species—have not been proved.

The frustrating gaps and inconsistencies in the record are as nothing compared with the surprising, ingenious discoveries and inferences made by molecular geneticists, anthropologists, and paleontologists in the previous 100 years. If there are huge questions, there continue to be confirmations of the general picture. For example, there are greater genetic differences detectable between separate African cultural groups today than between Africans as a whole and the rest of *H. sapiens.* Such diversity indicates that these indigenous peoples are closer to the physical site of the ancient birth of humanity, their genes closer to the evolution of our species.

Such confirmations do not, and may never, answer the questions of why species fail and related species appear. If chance plays the essential role, then science will never be able to address such issues. Science does not explain chance. Science seeks to find explanations for events that occur more than once and follow certain ascertainable rules.

As far as we know, a species does not spontaneously generate in two or more places, or at two or more periods in the evolution of life. *Homo sapiens* appeared in a certain place and time—and most certainly has only one opportunity to survive as a distinct species. Like our hominid predecessors, if we die out, we're toast. Extinction is a one-time thing.

In the constant rush of genes toward their future, certain packages of life, whether lemurs or human beings, carry DNA and other genetic material for a brief span on Earth. As part of that process, all life has evolved from the first living cells that somehow formed almost 4 billion years ago. The branches spread out; some break off, while others branch and branch yet again. Our species can be seen to dominate the Earth in recent centuries, but that image may be deceptive. Did the billions of years of galaxy formation, the 4.5 billion years of rumbling geology since the formation of Earth, at least 5 million years of hominid growth and experimentation and development occur only so that human beings could exist? There is a respectable scientific theory, the anthropic principle, that argues just that. Perhaps so—until the next punctuation mark.

9

Turing and the Brain as Computer, and Vice Versa

Intelligence. We are certain that it distinguishes us from the lesser animals; perhaps it is the principal reason for our physical conquest of Earth (begging the question of how long it will continue). Intelligence makes us, as far as we know, the single species that has a sense of history, probing ever more enthusiastically into a lengthy past that was not written down.

But the twentieth century revealed that human intelligence may not be, or long remain, solely the property of human beings. Can computers, which now suck up 6 percent of the U.S. energy supply, learn to think as capably and adaptably and creatively as the first group of *H. sapiens?* Are they actually machines at all when "artificial intelligence" is produced with organic materials?

In a different vein, is intelligence working optimally when it is precise, or should it be "fuzzy," must it be "fuzzy," in order to deal with perceived reality?

Almost as desperately as Victorian Europeans found reasons why their ancestors could not possibly have arisen in Africa, we have continually found distinctions between the thinking of humans and of machines. We allowed them to compute faster and faster, but that seemed to be gruntwork. We made excuses when IBM's "Big Blue" was able, it seemed, to outthink—that is, outreason—a chess genius. In short, many of these distinctions fell in the twentieth century. There may be a trend here.

As everyone knows, René Descartes thought he proved quite a lot when he announced "I think, therefore I am." But it may very well be true that an entity with the capacity to think as well as or better than any of us does not, in the sense the French mathematician and philosopher intended, have a separate existence; it may think but not be.

When movie star Tom Hanks explains on a tabloid TV program that the opening of a new film is "digital"—either it succeeds or fails, no in between, "it's 1 or 0"—and his comment isn't cut in the editing process, the assumption must be that the average viewer understands the basics of the "digital revolution."

If that's true, millions of us have come to understand, accept, and probably take for granted a mechanized "thinking" that was considered puzzling by many and disturbingly simplistic by some fewer than 50 years ago.

In the first place, the desire to create machines that could accurately store data, tabulate it, and perform computations with whiz-bang speed produced the first computers. Large, unreliable, and slow by our standards, they worked on the same principle enunciated by thespian Hanks: "It's 1 or 0." It was a revelation that wide areas of human thought could be mimicked by inorganic devices based upon that principle.

In the second place, amazement at the rapidly increasing efficiency and broader grasp of these devices inspired a related debate: If they could compute so brilliantly, could they in any sense actually "think"? Were they, or could they become, conscious? What Descartes meant was that his ability to think meant that he was a conscious being; in his brand of dualism, mind was separate from brain. Were the machines really capable of "artificial intelligence"? Not if consciousness is something separate from the hardware of the brain's circuits. To make this distinction between machine and brain is to distinguish also between computational skills, however rapid and comprehensive, and whatever it is that is satisfied to say, with Descartes, "I think, therefore I am."

Is consciousness, to refine the term, found outside the human brain in the natural world, in the plant you might have spoken warmly to this morning as you watered it, or in the dog who seems very interested when you head for the door? If the beast jumps off the couch when you shout "Bad dog!," does he understand the concept "Get off the couch," much less your comment about his wickedness, in the same way you do? Or is his action merely the result of rote learning? Most of us would probably designate all of our ape cousins as having consciousness, and perhaps our wiser pets, but not toads or goldfish or the deer in the woods. In discussions about consciousness, experts also talk about the possibility of zombies—that is, if an exact copy of a conscious human being could be manufactured, would it be conscious, too? And how would you know, since it would simulate all activities of its template?

These questions, of course, have not been answered to anyone's satisfaction, but the possibility of computer consciousness, because of structural similarities with the brain and its processes, has been tantalizing thinkers for more than a century. But this debate about the possibilities of "artificial intelligence," about the potential of computers to feel your pain or to learn that revenge is a dish best eaten cold, could not even find legs until the first machines proved that thought is a thing of 1s and 0s.

Jay
Forrester

In the early 1940s, when many a sword that would become a plowshare was forged, MIT grad student Jay Forrester was given the task of helping the U.S. Navy produce a flight simulator. This device, invaluable for saving pilots' lives and expensive aircraft by providing flight training in a mock-up on the ground, is only as useful as its ability to simulate the real experience as closely as possible. In simulators of the World War II generation, hundreds of mathematical equations simultaneously governed the device's nearly instantaneous reactions to pilot actions while also reacting to the imagined relevant forces of gravity, air, and engines upon any maneuver. The simulator was, in effect, a computer that sent orders to electromechanical gears and cams.

Forrester would not invent an improved simulator, but his work did lead to the creation, still backed by military funds, of a breakthrough electronic computer known as Whirlwind. For this machine he junked the base of 10 used in the decimal system in favor of binary arithmetic—that is, a system based on only two numbers, 1 and 0. There is, of course, nothing sacred or even especially efficient about the decimal system, except that it reflects the typical complement of human fingers. The ancient Babylonians did quite well with a system based on 12, still alive in our system of timekeeping and in the number of degrees in a circle.

For the computer age, a binary system was more suitable. In fact, the idea that all human thought can be conveyed with a series of yesses and noes goes back to British mathematician George Boole, who in the mid-nineteenth century set down "the laws of thought." In short, he proposed that any idea, no matter how complicated and profound, could be viewed as either 1, meaning "true" or "everything," or 0, meaning "false" or "nothing."

"What we cannot speak about," wrote Ludwig Wittgenstein, the philosopher whose work influenced Godel and most thinkers in the twentieth century, "we must pass over in silence." This is the key statement from his famous *Tractatus Logico-Philosophicus,* published in 1921. Some things, he meant, cannot be described in language; they can only be pictured. The pictures in our brain must match the pictures in the outside world if we are to make sense.

In the 1940s John von Neumann was influenced by this concept when considering how computers could be made to reflect reality in a stored program, the essential computer tool he is credited with inventing. He also showed that binary logic could be combined with arithmetic to provide an internal memory for a computer, its storehouse of data, and the instructions for dealing with data. By clicking on or off, an electronic device represents 1s or 0s; the speedier the device, the more extensive the information processed, stored, or passed onward. In this way, the formal math can communicate the most complex of human thoughts.

Translating a base 10 number to base 2 is very simple, however peculiar it might look at first. Take this example: 110101. Beginning at the right with the number 1, the value of each number is doubled by place when moving to the left: 1 + 0 + 4 + 0 + 16 + 32, or 53. To take another example, 10000 = 16, or 0 + 0 + 0 + 0 + 16. If this seems unwieldy to you, you already know from experience that it has proved to be tremendously successful.

Forrester's system would not gain acceptance right away, but increasing sophistication of hardware eventually made possible a computerized air defense system and landings on the Moon and Mars while moving out into our bank accounts, phone bills, and home entertainment centers. But the basic concept has not changed. When you are deciding whether or not to risk some retirement monies in a hot new over-the-counter stock associated with microchips, you are betting that smaller, cheaper chips will perform more efficiently and quickly the kinds of information sharing that, back in the 1940s, required thousands of fragile, easily overheated vacuum tubes.

But are the computers becoming smarter, to reflect advertising slogans, or are they merely becoming more adaptable and versatile? And what's the difference?

The work of Alan Turing, an English mathematician, was essential to the founding of the Computer Age. During World War II he was considered a national hero for inventing a decoder that could break the German enemy's famous Enigma machines, thereby giving the Allies a tremendous strategic advantage. But it was his idea of a computer called the Universal Turing Machine, developed in 1936 but never produced as a working electronic device before his death in 1954, that laid the conceptual groundwork for every digital computer in use around the world today. He also proposed, in the manner of Godel, his famous "halting problem." In effect, he proved that it would be impossible to predict when or why a computer might "halt" its operations—that is, fulfill its assigned task. Some operations might run on forever, searching for solutions, but no program could be designed that would vet all possible programs to determine whether or when they would "halt."

Apparently convinced that the machines he envisioned would quite quickly develop consciousness, he proposed in 1950 a kind of mind experiment to mark the arrival of that day, now famous as the Turing test.

Turing's example used a hidden computer, although you could perform the same test using e-mail. How do you determine whether or not the messages during a chat on your laptop screen are from a person? If your questions to your unseen interlocutor always receive replies that are germane and sane, is that proof that you are not talking to a machine? According to Turing, a machine that convinced you it was human would have to have human consciousness. It would not only be communicating on your level, it also would be thinking on your level.

The assumption here is that our consciousness is a product of the brain. Copy the parts of the human brain, or only the networks of specialized brain cells that play a role in consciousness, and you have replicated consciousness.

No machine has yet passed Turing's test, but if one ever does, not everyone will grant it human consciousness. To one school of thought, the computer will simply have learned how to simulate human consciousness by organizing information in a certain way, much as computers can simulate missile strikes on Baghdad or the face you will have after rhinoplasty. Simulation is not the real thing.

To prove the point, the philosopher John Searle came up with another thought experiment, the Chinese room. In a sense, it upends the Turing test, for the human in this case is hidden from view and provides responses that are tested for coherence and meaning. In fact, our subject sits in a closed room holding a large volume containing reams of strange symbols that are incomprehensible to him. When a piece of paper is slipped under the door, he is to look up its sequence of symbols in the tome, then write down on another piece of paper another sequence of symbols, as indicated, and pass this predetermined response back outside.

Little does he know, since he has never seen Chinese ideographs, that he is carrying on a very sensible conversation in high Mandarin. Good stuff in, good stuff out: he is performing the functions of an efficient computer and has been given the database necessary to do the job. But if the conversation has been about the best place to get Beijing duck, his saliva glands aren't watering; he has no idea what is being communicated.

In other words, the man has passed a version of the Turing test. It looks as if the unseen person/computer was thinking in Chinese when in fact he was not thinking—in terms of the Chinese language, not acting consciously—but following a set of instructions.

Nonetheless, the strange lure of creating a thinking machine, of creating a device capable of artificial intelligence, captured computer freaks from the beginning. The Perceptron, built in 1954, was a high-tech sensation of its day, attracting theorists and investors. Its photodetectors could recognize individual letters; so would it prove capable of "reading"?

As it happened, this particular attempt did not succeed, despite rewiring and other improvements, but it anticipated devices today that can scan a printed text and "read it aloud," as well as gradually improving voice recognition devices. These are remarkable achievements, but the bizarre, often hilarious corruptions that occur when you first use your new voice-recognition programming for dictation recall the lesson of the Chinese room. Unless these devices have a very clever sense of humor, they do not in any sense know what they are saying. At the moment, the best voice-recognition software is about 80 percent accurate and cannot distinguish between your voice and someone else's.

An amusing version of machine intelligence was developed in the 1960s by Stanford psychiatrist Kenneth Colby, who specialized in such personality disorders as paranoia. His program simulated a personality obsessed with the Italian underworld, but the same idea could be developed with reference to any obsession:

> Why are you at the grocery store?
> I'M NOT BUYING FOOD, ONLY TOOTHPASTE.
>
> Did you drive your car here?
> I EXERCISED EARLIER TODAY AND DIDN'T NEED THE WALK.
>
> Have you seen any friends here?
> EVERYONE SAYS I LOOK VERY GOOD.
>
> I hear this store might lose its lease.
> I'VE LOST 12 POUNDS IN 2 WEEKS.

And so forth. Each response is triggered by a word or a concept in the question, but only by a certain demented kind of reasoning. For a time, this machine might be able to pass the Turing test, especially with an observer who has spent much time on the commuter train to Manhattan and been forced to overhear one-sided cell phone chatter. But the cat is soon out of the bag, no matter how artfully and realistically such programs are designed. Current research may be getting closer to developing programs that can convincingly represent various emotions, but this is definitely work in progress.

Nor can the heralded success of Deep Blue, the computer that bested chess grand master Gary Kasparov two games to one (with three ties) in 1997, pass Turing's test of 47 years before. The computer's manual of Chinese, as it were, included the goals and movement options possible in a freewheeling game; it also was programmed to detect an opponent's predictable patterns of play or other possible competitive weaknesses. Even these latter functions, which are relatively complex, are no more than functions. Deep Blue could speak chess very well. (For comparison, a computer had bested the world's best backgammon player about two decades before.) In much the same manner, other examples of machine intelligence are now able to speak credit card (i.e., run the numbers to pick out credit card fraud) or speak medicine, crunching data to diagnose and recommend treatment for specifically defined medical problems. This is speaking, not thinking.

As for Turing himself, his honesty about his homosexuality, illegal in Britain during his lifetime and perhaps considered by government officials a security risk, led to his being incarcerated for a year and forced to undergo "organotherapy" to "cure" his sexual orientation. This therapy, an experimental course of drug treatments, had disturbing physical effects, including impotence.

A year after emerging from this experience, he was found lying in his bed dead at age 42, apparently having eaten an apple laced with cyanide. A coroner's inquest ruled his death a suicide, but his friends and supporters speculated darkly about the possibility of murder by agents of his own country.

Yet while computers may not soon write original verse drama based upon their own experience of what it is to think and feel like a computer, the way in which they operate suggested a new approach to understanding human consciousness in the twentieth century. Since we know how they compute, have we learned how we think?

Descartes was not the only one in history to suspect or hope that brain and mind are separate. In the Middle Ages it was thought that a *homunculus,* a complete and very small human being, sat inside the head producing thought. In the twentieth century many philosophers argued that something apart from, if within, the physical brain was responsible for human consciousness—an idea ridiculed as "the ghost in the machine" by Gilbert Ryle.

But if there is no ghost or homunculus, then is everything that distinguishes us from the beasts of the field to be found on electrochemical pathways in the brain?

Many brain researchers have decided that the answer must be yes. And it turns out that, at least metaphorically, the binary arithmetic used for the computer's calculations is similar to the electrochemical signaling that accounts for the mechanical processes of the brain: off/on, off/on, off/on.

By this imaging, there is no discernible difference—and never will be—between mind and brain. You are your brain cells, which include 100 billion to 1 trillion neurons and the glial cells that support their efforts in a mass weighing about 3½ pounds. What is missing in this picture? You could use the word "soul," if you like, or "free will," but the brain researcher will turn aside, at least during lab hours. The brain is all.

Like the inorganic computer at your desk, it stores information, learns algorithms (i.e., the procedures required to react to specific input in order to produce the appropriate output), and can take actions. Each neuron is equipped to receive and send messages—remember, we're talking about a group of cells that numbers as high as the number of stars in a midsize galaxy in the expanding universe. At the same time, they are highly specialized; the act of breathing, hardly a frill, is controlled by perhaps fewer than fifty brain cells.

Typically, a neuron sends out a message to another neuron down a single thin pathway called an axon. The content of that message depends upon information streaming in from thicker pathways called dendrites, which branch out plentifully from the neuron in many new directions. At each branching off, the dendrites can make contact with another brain cell, thus sending news from abroad back to the home cell. Nerve impulses are carried by substances called neurotransmitters that diffuse across a gap known as the synapse. Axons make the same type of contact from the home cell to other neurons when messages are sent outward. The average neuron, depending upon its specific function in the brain's neural networks, can make 100 to 10,000 or so synapses with other brain cells.

These brain cells, remember, were produced by decisions made by your DNA, and they were manufactured in the same way, except for one sad exception to the rule: unlike the cells that continue dividing as your toenails grow or a cancer spreads, neurons do not reproduce. That is bad enough, perhaps, but there's more: just by getting up, going through the day, going to sleep, and getting up again the next morning, we all lose an estimated 200,000 brain cells every 24 hours. Traditionally, these lost cells were considered just that, because brain cells, unlike other bodily cells, were thought incapable of neurogenesis, or being replaced. (Brain cancers, of course, require cell division, but they involve only the glial cells that act as a structural framework for the neurons.)

Beginning in 1998, however, published research began to suggest that at least two parts of the brain, the olfactory bulb and the hippocampus, can create additional neurons, even in elderly people. So might other parts, but it is not yet clear what functions the new cells perform or whether or not they can develop connections with neighboring cells.

With so many billions and billions of brain cells still on the job, do these daily losses matter much? Apparently that depends on the individual. Aging, despite our fears, is not necessarily tied to loss of memory and other brain functions, at least not progressively at the same rates for everyone. Genetic legacy may play a role, but the areas of mystery are yet as comparatively vast and dark as those bubblelike voids in space.

And still mysterious is the brain's ability to process and communicate the complex instructions required for you to hold this page, read it, remain sitting upright or awake on the pillow, be aware of your physical and emotional and intellectual reactions at lightning speed—and take action, if necessary, upon them.

Each of those information-hungry neurons receives and regurgitates messages through electrochemical pulses. Ordinarily, the cell has a tiny so-called resting voltage, or its characteristic electric charge; a specific chemical released at a synapse because of input from another cell changes the electrical charge at that point. The change in voltage lasts for a millionth of a second; in the same flicker of time, the cell's electrical charge returns to normal. If and when these input messages repeat frequently, the neuron responds by outputting a message down its axon: Get cracking!

The stove is hot, the finger leaps off. (And if you're good in the kitchen, you consciously decide to pinch your earlobe, the chef's way of getting immediate relief in such situations.)

Even this reaction response is extremely sophisticated, but what about the actions required to drive a car, especially if you're hunkered down in an MG surrounded by a pack of SUVs? All of your senses focus, or they should, on information coming in, or potentially coming in, from uncountable sources. To some of this information you will have to react, perhaps "instinctively." And your reactions will involve complex decision making, assessments, and neuromuscular actions, and perhaps a bit of human speech. The flight simulators of today are astonishingly realistic, but they have nowhere the capabilities you need in reserve just to buy a bottle of milk.

You can see how the structure of the brain and its basic on-off operation would inspire the metaphor that it is a sophisticated organic computer. You can see that it is reasonable to argue that further research should be able to find the mechanisms for every human thought and action. Is fear, say, caused by stored information that in the appropriate situation informs neurons to incite fear-oriented neurons, or does it occur because information causes neurons to activate such behaviors as a racing pulse that are associated with fear? Either way, it is all happening in the brain/computer.

By that reasoning, the emotions you feel are the consequence of your situation as perceived physically by a brain that has evolved over the aeons to keep you—or perhaps only your supply of DNA—alive and well. Some structures in the brain have been tied to an understanding of musical harmony; experiments with brain stimulation have shown that certain emotions or skills may be seated in specific, predictable sections of the brain. Drugs that are used for recreation and for psychological treatment work because they act upon the electrochemical activity of the brain. Some destructive behaviors, such as obsessive-compulsive disorder, have been cured or alleviated by stimulation of parts of the brain.

If these examples of mind-as-machine are of any interest to you, that is because your circuits are busy firing away, processing the information, then matching it against previous information (even if it could perhaps be called belief or conviction), and then deciding how to install it in your neural pathways—dump it, put it in holding position, or integrate it. This is the materialist view of the mind.

But you know that your brain is doing all those things. Therefore, are you not conscious? For the record, no one has a provable answer to that question; materialists and dualists are still working on it. There is even the "mysterian" school of thought: in effect, we are not equipped to understand the nature of consciousness, and we probably never will be. Perhaps. Or perhaps this century will be marked by unexpected forays into this compelling territory.

Back to the mechanical computers. If our brain is a computer but thinks it has a mind, what about "artificial intelligence" in that regard? Where would the difference lie?

One difference that has been addressed skillfully is logic. To better simulate human thinking, the machine has to be built to incorporate "fuzzy logic," which is essential to human thinking. This is not a term for bad logic but for logic that deals with the imprecisions of real-life experience.

For example, you could easily follow your parent's advice, if you wanted to, to wear warm clothing when it's cold outside and likely to get nastier. Rocket science not required. On the other hand, a very sophisticated machine designed according to the rules of classical logic could not process this advice, which is based upon human understanding that categories in the real world do not have precise boundaries. What, for example, is the color "red"? There is a very wide range of hues that could be called red by most observers in daily life. We have no problem dealing with the gradations, which fill pages in manuals of design specs but would freeze my laptop.

Classical logic is based on precise rules. The computer operator would have to input a precise temperature, or a value within a precise range of temperatures, for the concept "it's cold." Every other element of this parental advice would have to be entered with precision as well, for the machine to make a decision to act.

In other words, the neurons communicating with each other in your brain use fuzzy logic to understand the vague, imprecise, but very real concepts "cold," "warmth," and "nasty." Without fuzzy logic, our brains would experience continual bouts of vaporlock. Fuzzy logic requires judgment. When a machine is programmed with fuzzy sets, a concept can be represented by 1, and its opposite by 0. For example, unmistakable cold would be 1, and unmistakable warmth would be 0. Between the two extremes are the various possible degrees of coldness.

Machines capable of fuzzy logic were manufactured with great success beginning in the 1980s; by weighting data input, they are able to facilitate many different kinds of industrial production without getting hung up on inessential details. In this way they simulate human reasoning quite well, yet they cannot be said to make judgments. We carbonaceous life forms are still doing something, still have something, that our silicon-based computers do not.

One difference, you say, is irrationality. The computers have learned fuzzy logic but not the illogic that made a friend frantic the other night when she realized she had only half an hour left to drive to a drugstore that sells tickets for the New York State lottery. "You can't win if you don't play," she said with a laugh, tearing down the road. Did she really *think* she was going to lose $25 million if the store closed before she got there? What part of her neural circuitry would make that calculation? Or was she responding to something else, some sense of play or risk hard-wired into the brain when hominids were teasing saber-toothed tigers? Whatever, the computer doesn't have it yet. The computer would understand, however, why someone would call a government-sponsored lottery "a stupidity tax." The computer would be right, given its programs, but those who buy lottery tickets are probably right, too, and without having to listen to some inner homunculus.

Theorists of artificial intelligence, sometimes noted simply as AI, developed in the 1950s and 1960s a new and tantalizing branch of science known as cognitive science. The forerunner of this pursuit was *Cybernetics,* a book published by Norbert Wiener in 1948. Essentially he suggested that there is no difference between the computations of a computer and the thinking strategies of a human brain. Equally important to the science, whose adherents consider it a "revolution," was Noam Chomsky's *Syntactic Structures,* published in 1953. He argued that a kind of "universal grammar" for language, passed down genetically through the centuries, exists in the human brain at birth. This grammar and other inherited structures, according to cognitive scientists, are evolutionary developments designed to interact constantly with the outside world. The child learns its native language, whether English or Wolof, but is born with the basic concepts of language, as with the structure of reason. Chomsky's image is an intricate system of wiring in the neonate brain that is not completely hooked up. Its switches, to continue the metaphor, can be set in only a few specific ways; it is the child's exposure to the language of the hearth that determines exactly how the switches will be set.

In much the same way, according to cognitive scientist Gerald Edelman, the brain's pathways exist at birth in a predetermined pattern. The human's experiences with the world outside, however, will stimulate the brain to choose certain combinations of neural connections. The individual, using the same equipment provided to everyone else at birth, creates mental maps that categorize the world in an individual way, changing the maps as life provides new and different experiences from outside. The brain takes its specific shape in order to deal with its specific set of experiences, but its essential structures, though flexible, are inherited.

Cognitive science, in short, uses the model of the computer for the brain in order to study the nature of human knowledge. Bringing together such diverse sciences as linguistics and AI, psychology and linguistics, cognitive scientists remain wary about the limitations of their study. Emotions, they say, are too vague and complex to be analyzed by their science. Still, by examining human thought as a computational process involving symbols manipulated within given structures, they reinforce the idea that there is no homunculus lurking between the neural pathways. We are what we inherit—brain cells—and what we experience—the brain responding to the environment in pursuit of its goals, or programs.

Cognitive scientists agree, in short, that brain and computer use similar symbolic systems, and that the latter is an effective tool for analysis of the former. One scientist likens consciousness to a computer's operating system. Continuing that image, consciousness stores files and, when necessary, alerts the brain when there is a problem, such as an unformatted or full-up diskette. Not to imitate the brain but to understand its workings, computers have been developed that, to some extent, perform humanlike tasks, such as learning how to turn English verbs into the past tense or to imitate the finger movements used in typing. No one, as far as I know, suggests yet that such experiments are about to trap the ghost in the machine.

The mind has been developed to cope with what it needs to know; given that assumption, it is possible that the ghost—that is, mind as opposed to brain—is not, for our survival, on a "need to know" basis; it could be there, but we don't need to know about it in order to survive. Some who ponder the possible nature of the matter missing in the universe, the so far undetected 95 percent of mass that must be there, speculate that it is right in the room with us, but our brains have no need to know about it. Close your eyes quickly. Then open them, seeing what you can see. In the jumble on your desk, in a corner of your living room, you cannot possibly count the kinds of information you have just perceived in an instant: colors, depths, shapes, distances, light, dimensions, relationships, textures, categories precise and fuzzy. Look at how many bits of information are necessary just to look at a pile of magazines, a row of books, the bottles shelved behind the bar. All of it is there instantly, regarded and analyzed and taken for granted. That is the organic computer at work. But as TV interviewers put it, "How do you feel?"

Will even the accelerating development of computer capacities ever bring the machines very close to what we achieve every instant without thinking about it? In 2001, the market's best available microprocessor uses some 42 million transistors to execute up to 1.7 billion commands per second. Soon, according to researchers, silicon transistors only 3 atoms in thickness and 70 to 80 atoms wide will be available. At this size, quantum mechanics becomes an issue, as it does even more so in current experiments to create quantum computers that use individual atoms as transistors. High-frequency radio waves are beamed at the atoms, forcing them either to point upward, indicating a binary value of 1, or downward, indicating 0. According to quantum mechanics, each atom can be in both positions at the same time; this doubling theoretically means that a computer using only 40 atoms could simultaneously execute 10 trillion instructions. Would this vast informational capability eventually produce something that we could recognize as an intellectual relative?

The cognitive scientists, to repeat, have cautiously excluded the emotions and all the messy parts of personality from their investigations, but it is surely difficult for us to read about their work or any advances in computer science without wondering whether or not consciousness, as understood for millennia, may not be an illusion.

Do I want to live so that I can enjoy the company of everyone I love as long as possible? That's what "I think." But perhaps my trillion brain cells in their nearly infinite wisdom are simply programmed for survival and all of the structures/connections have evolved to keep "me" focused and fairly happy while I perform that function for the species. We thought we knew who we were, and *that* we were, before we got to thinking too much about computers: what they do, how they do it, how they resemble our brains. But I will not be I, we now know, if injury, drugs, a tumor, or disease affects the mechanism that thinks, and that believes it is able to think about thinking.

In the Alzheimer's disease ward where my mother has lived for some years without knowing it, there are many different kinds of behaviors. Some of her neighbors sleep through the day, some sing hymns or love songs over and over, some curse, some weep. Her roommate, once a very gentle and gracious and witty woman, is wide-eyed with some nameless dread, unable to make sentences, convinced that her relatives are being buried at night in the parking lot outside her window. She hugs her stuffed black dog until the thin muscles in her arms tremble. What has happened to her, perhaps because of the instructions given by one or more of the three genes that have so far been implicated in various forms of the disease, is a strangling of various brain communication points by an amyloid sheath. Evidently it grows indiscriminately through the galaxy of neurons, destroying or interfering with this person's ability to recognize family, that person's sense of trust. What a strange lottery this is. My mother, whose cognitive deficits are severe, who cannot speak or feed herself or reason in any communicable way, is bright-eyed and good-humored. She smiles when she sees a child or a pet, as she always did. Those connections have not been cut off by the creeping tentacles of the disease. Is her consciousness hidden somewhere behind barriers, or has it died in pieces with the dying of her brain? She was terrified and bitterly angry as the first circuits closed down, the plaque first began forming, but obviously she is serene now; the brain parts left must experience life as benign. And she is conscious. So we believe.

Not only might it be unsettling to consider our private thoughts, our insistent desires, our great mental pleasures and treasures as completely a matter of electrochemical action down neural pathways—no ghost, no homunculus—the idea is also a gauntlet flung down against many branches of psychology.

Is the analysis of hidden fears, repressed emotions, and deep-seated emotional traumas irrelevant to the brain's computer? If such things affect the roiling activity of a conscious, well-functioning brain, how do they do so?

10

Freud, the Unconscious, and Other Views

Cut out a part of the brain, it turns out, and the subject may suddenly have a very different personality. The meek may turn wild, the angry quiescent.

So . . . the person we knew before such an operation was not the real person? The changed person is? Or is there a personality that included both versions? Phrase the question any way you like, it would seem that there is a strong connection between the composition of the brain and the behavior of its owner. Throughout human history we have, it's pretty clear, held individuals responsible for their behavior. We have perceived the lowly in the same fashion as famous epithets have crystallized our perceptions of the mighty—Edward the Confessor, Bloody Mary, Harry the Wise: each person is a distinct personality . . . until we spoon out a portion of the brain. Then Harry is no longer sage, and Mary goes to Calcutta to empty bedpans. In other words, we may have found out that a human being is indistinguishable from the configuration of his or her brain. My brain thinks, therefore it is?

Then where am I? And what?

If studies of the brain cannot answer such questions to the satisfaction of many, what about the numerous attempts throughout the twentieth century to assess and explain individual personality?

"Sneezing is a sign of rejection." "He is terrified of intimacy and commitment." "She's a walking neurosis on the verge of psychosis." "Those two are together only because they're both driven by repetitive traumas." "The man's testimony was one long Freudian slip." These are the kinds of comments we all hear or make without a second thought. And no Beatles fan ever had to ask what John Lennon meant by writing "I've had enough of reading things by neurotic psychotic pig-headed politicians."

But whether in Woody Allen movies or over martinis, the clichés of analysis are not only well worn, they also may reinforce the notion that only one kind of study of personality, whether healthy or disordered, prevailed throughout the twentieth century. In fact, psychology from Freud to the present has been a complex journey marked by frequent, often angry disputes between advocates of different ways of attempting to understand what produces the individual personality, and how it does so.

Putting the question of consciousness aside, it can be agreed that personality, though equally impossible to find and dissect within the body, is the sum of an individual's behaviors. At least that is our personality as experienced by others. Hidden aspects of personality are important to an observer only to the extent that they are revealed or suggested by actions.

As long ago as ancient Greece, and probably since we first appeared in East Africa, humans have recognized that there are distinct personality types. Surely there always have been words for the concepts we tend to describe as "depressed," for example, or "aggressive." In many cultures personality types were associated with parts of the body. To take one example, the idea survives in the word "sanguine," the derivation of which links the blood with a robust physical or emotional makeup. Physicality, to repeat, and personality were interrelated, probably the one determining the other.

In the mid-nineteenth century, partly in response to the supposed increasing strains of urban life in Western Europe and the United States, many researchers looked for physical explanations for conditions that we would now view quite differently, thanks to the warring schools of psychology in the twentieth century.

In the 1860s, for example, the New York City physician George Beard drew upon personal experience to create a theory covering almost all mental problems. Before choosing to go to medical school, young Beard had suffered what we might call severe anxiety attacks verging on nervous breakdown. He might even have been struck with the kind of schizophrenia that can sometimes affect people in their late teens or early twenties, then spontaneously disappear. In any event, once he had established a practice, he began to notice that many of his patients seemed to be suffering similar mental and emotional disabilities. With no exceptions aside from outright raving madness, Beard decided, all of these problems could be grouped together as "neurasthenia." The cause? The body has a finite reserve of energy stored in the nerve cells, and it was being exhausted quickly by the pace of "modern" life, leaving his patients confused or stressed out or even delusional.

"The patient may be likened to a bank," agreed Boston's Dr. J. S. Green, "whose reserve has been dangerously reduced and which must contract its business until its reserve is made good." Some experts feared that the flood of nervous disorders might prove Charles Darwin all too correct: Humans might not evolve quickly enough to survive the changed environment of industrialization. Women were considered especially vulnerable because they must have smaller supplies of energy and, in addition, were drained by the demands of their reproductive organs. They were especially prone to hysteria, typed as a "woman's disease," which was supposedly provoked by a uterus moving out of position inside the body. But men and women alike were succumbing in greater numbers to neurasthenia in some form, according to alarmed physicians.

Much quackery was inspired by these fears, from electrotherapy—stimulating the brain cells and nerve fibers with low-voltage forms of electrical current—to the promotion of drinks with allegedly restorative chemicals or stimulants, including Coca-Cola. The important factor is the underlying assumption that restored nerve cells would promote mental health. Emotional stability was a matter of healthy neurons. *Mens sana in corpore sano.* (A healthy mind in a healthy body.) And a nice cuppa will have you feeling better in no time.

But the quackery did not produce cures. Nervous conditions seemed to be proliferating.

Now, if the emotions are affected by the body, it might seem logical that heredity would play a major role in neurasthenia. Although the mechanism was not yet known, it was clear that physical characteristics blossomed again and again on the family tree: If hair color, why not nervous deficiencies?

In pursuit of this notion, the French neurologist Jean-Martin Charcot began an intensive study of the possible cause of hysteria in mental patients at a Paris hospital. Noting that some were male, he sensibly put aside the uterine theory and focused on the brain; in autopsies of the disturbed, he hoped to find a connection between lesions in specific parts of the brain and the late patient's characteristic pattern of tics, spasms, fits, wild gesturing, and other visible symptoms of hysteria. He assumed that weaknesses of the brain, of the nervous system, were inherited, including the defects inherited by parents and the scars they acquired over their lifetimes.

Charcot was tireless, became famous, and failed to find the answers he sought. Unwittingly, however, he would greatly influence the brilliant young Austrian who would completely overturn the day's approach to psychological disturbances.

A budding neurologist at the time, 29-year-old Sigmund Freud had concentrated his attention on the structure of the nervous system, especially the brain, and on the effects of drugs such as cocaine on specific nerves. He studied under Charcot on a 4-month-long fellowship in 1885–1886. On one point, the student was already better informed than the master, for Freud knew from lab experience that the structure of the brain was much more detailed and specific than Charcot yet understood. Freud also recognized that no relationship between hysteria and brain trauma was going to be found. What piqued his interest, however, was the older man's treatment of certain emotional problems with hypnosis.

Certain of Charcot's patients exhibited physical symptoms that made no physiological sense. A paralysis in an arm or a leg, say, would extend to regions that would not ordinarily be affected as well, considering the physiology of the nervous system. Evidently the cause was not physical but mental, and indeed, the patient under hypnosis might be induced

Sigmund Freud

to move the supposedly affected limb. "Hysteria," Freud wrote later, "behaves as though anatomy does not exist." Instead, he surmised, the mind was somehow overriding the user's manual for the body. And, at the same time, the patient had no idea that the mind was acting in this way. The decision to paralyze a limb was unconscious. Under hypnosis, patients would not only lose the mind-determined deformity but also would, when given the suggestion, mimic another one.

When he returned to his hometown of Vienna to set up shop, Freud did not immediately stray from the orthodox practice of neurology. As late as 1891, he wrote a neurological text on aphasia that is still cited by experts in the field.

Tentatively, he tried hypnosis on certain of his patients, who tended to be troubled young women from respectable, well-off households in the great and thriving city. By indirection at first, he was developing his concept of an unconscious mind, an area of human emotion that was powerful but mute. For many of his later critics, a key problem arose from the nature of his clientele. As an older man, was he something of a chauvinist, who encouraged or imagined sexual dysfunctions in these younger, attractive women because those data might reinforce his theories? Or were these frustrated women (as his defenders today suggest), stifled in bad marriages or potential spinsterhood, the products of a society that gave them no opportunities for creative development outside traditional roles in a bourgeois household?

Such questions implicitly suggest that the bright young theorist was experimenting for the ages, had intentionally cast his net in certain waters in order to practice his craft. The truth, according to the famed American neurologist Oliver Sacks, was very different; the empathetic doctor simply filled a treatment vacuum: "People did not come to Freud to be investigated; they came because they were tormented, because they were obsessed, because they were driven, because they were jealous, because they were frustrated, because they were depressed, because they were anxious, in some cases because they had strange symptoms which could not be explained by organic neurological disease. People came as patients to be helped."

Freud's hope for hypnotherapy was to discover what unconscious concerns might be causing the distress of his patients. Find out the cause lurking in the unconscious, Freud thought; then the mentally disturbed would be cured by dealing with it consciously. In fact, he discovered, a patient might indeed recall a horrific or puzzling experience in the past while under hypnosis but forget it entirely when she was brought out of the trance. In other words, her unhappiness could not be eased by discussing something she could not remember except under hypnosis.

Freud also found out under self-hypnosis that the unconscious could produce fantasies about events that never happened; in particular, he recalled situations or feelings that would later be described as Oedipal but, he felt, had never actually occurred in his own childhood.

Freud dropped hypnosis, not a reliable form of truth serum, for "the talking cure," keeping the patient conscious but relaxed. Yes, he did use the couch you have seen in a billion cartoons about analysis. Evidently a patient in this situation can "free associate"—babble aloud about whatever comes to mind. With the shrink out of sight—and therefore not providing a visible, distorting audience—the client lets it all hang out without fear, assuming that the whole scene is felt to be safe and supportive.

In theory, the free associating from word to word, mental picture to mental picture, will allow a person to wind around, skirt, come near, and eventually face head-on whatever is causing emotional problems. In theory also, this cause is hidden in the unconsciousness, where crippling anxieties are produced. Bringing out the forgotten, making the unconscious conscious, would produce healing. Freud believed, to quote the influential twentieth-century philosopher Michel Foucault, "in the possibility of a dialogue with unreason."

By 1896 Freud invented the term "psychoanalysis" to describe his work. Gradually, his continuing investigations would create the firm outlines of his legacy. As in the unconscious so in dreams, he came to believe, there were hidden clues to personality. The hidden emotional data might include repressed sexual fantasies and feelings, perhaps dating back to infancy. And, according to his theory of personality, we are divided, like Gaul, into three parts: our id, the punk in it all, is driven by unconscious impulses to gain immediate gratification; our prisspot ego consciously clamps down on those impulses as a matter of self-protection; and aid and comfort in that effort come from the superego, a grab bag of internalized curbs presented to each of us by our particular emotional environments, from family and friends to rabbis and sports heroes. According to this scheme, fears and desires war with reason, but many a battle takes place without our conscious knowledge, unreported to consciousness, so that the victor drives us to behave in ways we think we choose but actually do not.

These ideas were, and have remained, controversial. As Freud himself said shortly before his death in 1939, "I have discovered some important new facts about the unconscious and psychic life. I had to pay heavily for this bit of good luck."

Not surprisingly, it was the extraordinary notion of infant sexuality, which he first proposed in 1905 in his *Three Essays on the Theory of Sexuality,* that provoked the most serious attacks. In addition to the scandalous nature of such ideas—for example, that a child could feel sexual attraction for its mother as part of an Oedipal complex—there was the charge that his work was not science but mysticism. Pure hoodoo stuff, it could not be proved. It was not falsifiable. It was, by definition, subjective and therefore resistant to objective analysis.

If I don't believe that I have a "death wish," for example, your effort to prove me wrong will fall instantly into circular reasoning. If I refuse to believe that the spire of a Gothic cathedral rivets my attention because of "penis envy" rather than because of my yearning for the infinite, you cannot demonstrate otherwise—at least, to doubting me. (It was Freud, after all, who observed, "Sometimes a cigar is just a cigar.") These notions roost in the mists of unreason.

Nonetheless, the charismatic Freud began to collect a regular circle of followers in 1902; six years later, in a traditional emblem of success, there was an international congress of the newly formed Vienna Psychoanalytic Society, to be followed by the formation of the International Psychoanalytic Association. Soon there emerged further indications of success: Embittered and estranged disciples, including Alfred Adler and Carl Jung, founded their own schools of psychoanalysis, and others also would strike out on their own. The gospel spread in many different versions.

But what of hysteria? Ironically, it was not the moist recollections of young Viennese women that would bring Freudian ideas to the larger world but the horrific suffering of apparently uncomplicated young men in the trenches of the Great War, itself a convulsion of irrationality on an international scale.

"With staring eyes, violent tremors, a look of terror," according to one account, it was at first the French and Belgian combatants who responded as had no soldiers in epic battle ever before. While mortar shells rained down in dread arcs from German emplacements, roaring night or day in a drumming of noise and air pressure, boys and men ran out into the teeth of the chaos. They laughed, they wept. They were, it would seem, hysterical.

It soon became apparent, once the Allies brought up sufficient supplies to match the ferocity of the German invasion, that these symptoms were not confined to weak-kneed, neurasthenically inclined *poilus,* as Kaiser Wilhelm's officers had at first superciliously believed. Soon the Boches, too, had their own emotional wrecks to deal with during bombardments. "Shell shock," it was called. The term stuck, but it was not inclusive; men far removed from the shelling also fell suddenly mute, went deaf, or fell into fits of shaking and mumbling.

Eventually, in a reckoning that must surely be incomplete, about eighty thousand soldiers on both sides of the insane conflict were diagnosed with the condition. Some, like my paternal grandfather, would have periods of apparent lucidity throughout the rest of their lives, but never fully recover, never work, never become again husbands and fathers, so that my grandmother's suicide and the emotional problems of the next generation were, as we say now, collateral damage of war. Other veterans seemed to heal in the peace between the wars, and if they failed here and there, kind neighbors spoke of battle fatigue or "soldier's heart."

Still, young Johnny with the amputated hand was welcomed back as Johnny minus a hand. Shell-shocked Johnny was "never again the same." The constant of a defining personality has been forever altered. If Johnny had always been "reliable as clockwork" or "gentle as a breeze," perhaps he was now something else, someone else. Trauma catalyzed the mutation.

Since television dramas, not to mention talk shows, include "posttraumatic stress disorder" as part of the basic show-biz lexicon since the Vietnam conflict, the appearance of shell shock on the fields of France may seem like very old news. To military leaders of the time, however, another term of art was used: "cowardice." Indeed, many sufferers were executed as examples; but such rational therapy did not speak effectively to the irrationality of the disorder.

As physicians tried various futile treatments, psychoanalyst Freud got it right away. Much as the hysterical behaviors of Charcot's patients were based in the unconscious because of an inability to deal with a repressed memory, in a kind of turbocharged sequence of overloaded input the soldier afflicted with "traumatic war neurosis" was unable to deal with the irresolvable conflict of his situation in the trenches: prevented from charging across the field toward the enemy as the shells poured down, he was forced to squat immobile under the barrage, unable either to fight or to flee. This conflict was internalized, in the language of psychoanalysis, in the unconscious. Yes, the boy was conscious of the onslaught, but he put the essential "fight or flee" conflict down deep in the mind, or did not even recognize it in those terms.

The manifestations of shell shock, from tics to catatonia to weeping to whatever, spoke the language of impotence. These men were the casualties, as one neurophysician said at the time, "of nerve warfare." Ever afterward, as if from mental swords to a mental plowshare, many laypersons and professionals were willing to grant that Freud understood something new and fundamental about human behavior and the emotions. They saw the data for themselves on Main Street or at family reunions. Clichés of psychoanalysis entered the popular press. A dreadful type of misguided biographical writing, the psychobiography, would arise to titivate the naive. Once released throughout world culture, Freud's ideas became familiar to the point of caricature. On the other hand, some still could not accept the notion that personality had been riven by severe mental conflict. Some of my relatives, for example, insisted that my grandfather had been gassed.

As we know, where there is action, there is reaction—a truth in the sphere of ideas, as in Newton's eighteenth-century physics. Behaviorism, a major trend that achieved wide notice with the publication of psychologist J. B. Watson's book of the same name in 1925, chucked consciousness, the unconscious, and all subjectivity. Some 12 years before, Watson had begun spreading the essential ideas of this new psychological study among professionals. The goal was to focus on behaviors objectively, ignoring what could not be clearly observed and, if possible, accurately measured. This concept has been jokingly enshrined in academia, as a cognitive scientist in David Lodge's witty recent novel *Thinks . . .* explains: "Two behaviorist psychologists have sex, and afterwards one says to the other, 'It was good for you, how was it for me?'"

Watson and other behaviorists believed that the measurable behaviors were essentially the product of environmental factors. In laboratory experiments on humans and the lesser animals, researchers hoped to discover the general laws of behavior while admitting that the enterprise was a daunting challenge and would take a long time.

Again, certain aspects of an intriguing new approach to personality entered the public consciousness in fragments. The Russian physiologist Ivan Pavlov, immortalized in casual speech in the catchterm "Pavlovian response," began his career in the late nineteenth century by investigating how the brain communicates with the body's other organs. In one experiment, he operated on a dog so that food it swallowed would not reach its stomach. Nonetheless, the empty stomach produced gastric juices in order to digest the food. Pavlov correctly surmised that the canine brain, based upon input from the dog's mouth, sent orders for the stomach to get to work.

But it was his study of reflex action, begun in 1903, that captured the imagination of professionals; the public; and, given the nature of the Communist government that would rise to power, politicians intent on "brainwashing." Like you, a hungry dog in Pavlov's lab began to salivate when he glimpsed food. This reflex, in the argot of behaviorism, is unconditioned. Pavlov added a bell. It rang each time the dog was fed; before long, the tinkling alone caused the animal to salivate, chow unseen. The dog had associated the sound of the bell with the approach of food, causing it to develop a conditioned reflex. Commissars salivated.

Human experience, human development, and human possibilities are infinitely more complex, we like to think, than the cravings of a lab animal, but it seemed to behaviorists that we simply do not recognize how profoundly we are "conditioned" by our immediate surroundings. The increasingly messianic Watson boasted that he could take a group of children (perhaps it was a Freudian slip that caused him to choose the number 12) and train each one "to become any type of specialist I select—a doctor, lawyer, artist, merchant-chief, and yes, even into a beggar-man and thief, regardless of his talents, penchants, tendencies, abilities, vocations, and race of his ancestors." I become a survivalist, you become a Buddhist mendicant, in virtually the same way the pigeon in the Harvard lab learns to play the piano. His reward is grain; ours may be more sophisticated.

On the downside, perhaps, such a program threw free will out the window. Behaviorism could be read as strictly deterministic. On the upside, it also disputed the eugenicists and others who put stock in heredity factors as essential to the development of personality. Inferior races were not gumming up the American future; inferior environment was the culprit, shaping the individual personalities that collectively made a nation. Tweak that sentence just a bit more in the obvious direction and you see how behaviorism could be of interest to all kinds of well-meaning social engineers, not just Soviet commissars, and thus raise some disturbing ethical questions.

If we do not have minds but only patterns of behavior, and those patterns of behavior can be created or changed by laboratory experiments or other controlled stimuli, then we are, possibly, creatures so malleable as not to have anything most of us would consider to be a defining personality. At least, that is the specter raised by some in response to the images of lab animals and human volunteers running through mazes, pressing buttons on command, and otherwise submitting to conditioning.

On the other hand, behaviorism begat behavior therapy, a treatment complementary to Freudian psychoanalysis and generally accepted as benign. Of your own free will, whatever that might be, you might choose aversion therapy as a method of giving up a bad habit, such as smoking or watching reality TV. In some way, the undesirable behavior will be associated with a displeasing stimulus that is repeated until you are conditioned to abandon your bad habits. Much more commonly used, however, is the behavioral treatment for deep-seated fears. In desensitization, you will be taken again and again into the situation that causes your phobia. By definition, a phobia is irrational, out of proportion to the actual danger. With repeated contact, combined with relaxation techniques to relieve your anxiety, your fears will be conditioned to a manageable level. The success rates for aversion therapy and for desensitization are not undisputed and are perhaps greatly keyed to the attitude and goals of the patient. A convicted sexual offender sentenced to rote behavioral treatment is less likely to respond than an alcoholic horrified that she has blacked out and has no recall of driving her toddler home from kindergarten.

But some behaviorists, most famously B. F. Skinner, have suggested that our supposed intentions—indeed, any kind of consciousness—are irrelevant to behavior. "We're always controlled," he said, "and we're always manipulated." We certainly will be both if we happen to fall into a so-called Skinner box, his generally misunderstood invention to test operant conditioning, also known as instrumental conditioning. (In the early 1960s there were quite a few colorful rumors around Harvard Yard about Professor Skinner's raising his own daughter in such a box, but I take these tales to be urban legends.)

Reinforcement is used in operant conditioning to produce behaviors. Play the piano, get the seed corn, and so on. This systematic manipulation of the consequences of a behavior becomes, in a controlled experiment, also a means of predicting behavior. Put out the seed corn, hear the piano key strike. In the Skinner box the rat is closed off from the world. Buttons or levers are at hand. If pushing a button earns a reward, such as food or escape, the rat comes to believe that this action will always result in the same reward. The reward is the reinforcement of his new conviction.

Note that conscious thinking is not required to make the association. That is Skinner's point. The rodent's behavior can be understood by the experimenter without resorting to concepts of mental processes or any type of internal mediating. All is external. Whether we raise the ax to chop down the tree or sever Mrs. Borden's neck might depend upon our environmental conditioning.

Behaviorists are well aware that their theories can be ridiculed as reductionist. Many extravagant expectations for their approach to explaining or altering human behavior have been disappointed. Their lab experiments can be derided as bits and pieces of the variegated panoply of human experience. Prosecutors, and quite a few juries, are not generally happy with the defense that argues that a bad upbringing caused a defendant to perpetrate a criminal act.

Still, behaviorism is alive and well in universities, and Madison Avenue looks in from time to time.

It has been mediated, however, by the ideas of cognitive psychologists, whose analyses are today informed by cognitive science. In the early part of the twentieth century, however, they distinguished themselves from radical behaviorists in a very clear way. While the latter proposed only that the environment, natural or manipulated in the lab, changed behaviors, the cognitive psychologists believed that the change in behavior was connected to a change in knowledge about the environment. In other words, a subject might respond in one way to a stimulus, but his experience might cause him to change the way he responds in the future, now that he understands what the environment is capable of. To the cognitive psychologist, therefore, there are "knowledge processes" that influence how one responds to a stimulus. Behavior is not unthinking response, at least for humans.

For the disturbed patient, this kind of thinking—for that is the point, it is *thinking*—can be used in treatment. If you are depressed because every major stimulus in your environment causes you to respond with a pang of self-hatred or worthlessness, the cognitive therapist will help you think your way out of this cave. The trick is to learn to view stimuli as positive, then make these positive reactions as automatic as the negative ones, thereby knocking the latter out of your mind. Yes, mind. Cognitive therapy is an intellectual approach to emotional problems. You can use logic to change your feelings according to this approach. The behaviorist therapist, remember, would condition you, and the psychoanalyst would have you revisit the repressed, traumatic pain in order to deal with it.

In short, there's no agreement on the nature of personality, or its connection to consciousness, which has not been proved to exist. But, for mild diversion, let's note that such problems are not being left solely to the card-carrying psychologists these days. Somewhere out there, perhaps, is an explanation for consciousness that will bring in quantum physics, will bring Schrodinger's cat into the Skinner box. Remember that the quantum wave of possibility collapsed into one reality, dead cat or live cat, when the interior of the box was perceived. Put another way, it could be theorized that human consciousness must indeed exist because it is human consciousness, operating through certain parts of the brain, that eliminates alternatives in the quantum world by causing the wave function to collapse. (Consciousness doesn't choose which alternative will become reality according to this theory; it just forces the choice by observing the wave that is the range of possibilities.) Without human consciousness, without some spectator to register the condition of a confined cat, quantum waves of probability could not collapse.

Alternatively, according to some far-out thinkers, consciousness is not the cause but the manifestation of quantum collapse. In this view, the tiny cylindrical protein structures in the brain's neurons known as cytoskeletal microtubules, as noted previously, channel the collapse of quantum waves of probability; our subunits of consciousness are thus informed of these events.

I think . . . therefore I determine whether or not the cat in the box is alive. Not very satisfying to my personality.

Psychoanalysis and behaviorism were major twentieth-century approaches to personality. Psychosurgery and various drugs became important toward the end of the century in altering personality by working directly on the brain. Cognitive science, using the computer as both tool and metaphor, studied humans as thinking machines.

All of these approaches hold a part of the stage today; none of them, perhaps, gives succor to the hope that Descartes's formulation tells the story. Consciousness, behavior, personality, the unconscious—each of these ideas is as fluid as Heraclitus's stream. If someone promises to tell you who you are and how you got that way, check to see that your wallet's still on your person. We have lost, for the moment, any real hope of establishing the nature of self. We are afloat.

Instability is not necessarily disturbing, especially if the twentieth century discovered that it is undeniably a basic tenet of existence. What is, is. But only for an instant. By the time we grasp the present moment, it's gone, so we can never really grasp it. (Perhaps this is what Jacques Derrida meant by "an always already absent present." Perhaps not.)

The quantum world shows that physical existence is not predictable, much as Godel showed that systems of math would lack decidability, much as the Earth and its solar system and its galaxy are not fixed in a spot that can be measured except in relationship to vast structures speeding away at phenomenal speeds, much as your infant is not a foolproof recipe in a box.

Are you kind because you choose to be a kind person, or because of inherited brain structure, or because of behavior in your home environment? As your cells multiply and divide, following instructions given unconsciously by messages from your ancestors, are you only nominally a freethinking individual? Will a blow upside the head change your behavior from Jekyll's to Hyde's?

If by some miracle we come back around in 50 million years, we will see northern Africa jammed into Europe, obliterating the Mediterranean Sea. Aeons later, all the continents will agglutinate into one again after innumerable Ice Ages and magnetic reversals of the poles and a scattering of the shapes of our familiar constellations of the night. Constant flux and flow at every level of existence, possible barriers to final knowledge in many different fields—that is part of the story of contemporary science. In the 5 million or so years we likely have left as our form of hominid, if one rule of speciation holds, what can we do except get as much of the whole story as we can? Our brains—and perhaps our minds—crave it.

In technothriller author Philip Kerr's brilliant novel *The Second Angel,* a mysterious presence (no surprise endings given away here) speculates about the future: "There are no miracles except the science that is not already known. And man is the measure of all things."

In that measuring, which has become so ambiguous, we will never reach an end point; the music never stops. This is what we are learning, and this is what we will learn to celebrate. We and all we understand have no home but constant, rapid, accelerating change. "Everything fluid; random; potential," as Gore Vidal notes in his time-travel novel *The Smithsonian Institution.* "Like life itself."

Exactly a year after I began this book I am back in the converted barn on Long Island. There is barely a suspicion of autumn; and fresh ears of corn, truly remarkable this late in the waning of the year, are set out at the roadside stands, but few drivers are on the road. There are no cigarette boats in the bay nearby, no hammering from off-season construction or remodeling, and only scattered patrons in the shops and restaurants of the three blocks of sun-washed Main Street, for it is only a month after the collapse of the World Trade Center, and the world has changed immeasurably.

Back in the summer, when such things found room on the front page, scientists discovered evidence far off in space and time that might suggest that trusted universal constants, perhaps even the speed of light, may have varied since the big bang. Big bang theory itself is being strongly challenged for the first time by ekpyrotic theory, in which our nascent universe collided with another universe in a hidden dimension; the resulting big crunch jump-started our big bang. Mysteriously, and contrary to twentieth-century astrophysics, energy was detected streaming out of a black hole as massive as 10 million suns late in 2001. Meanwhile, a growing number of physicists, thinking counter to nearly a century of noodling about the quantum world, have begun to theorize that there must be an explanation of subatomic conundrums that would make sense in reality as we know it, reconciling quantum equations with classical physics even as other theorists now feel that they are coming very close to melding relativity and quantum theory at last.

We keep using our brains. For a time, we continue.

Acknowledgments

During the most difficult year of my life so far, as I dealt with a series of deaths and their innumerable numbing consequences, my editor, Jeff Golick, patient with me and thorough with my manuscript, dealt with the sweeter but exhausting consequences of change at the other end of life, the birth of his first child, a son. I am grateful that chance gave us this contrasting time, fortunate to be the beneficiary of his sharp eye and fusillade of queries.

Others who have helped this project directly and indirectly, and in some cases perhaps to their surprise, include David Schiff, David Kaiser, Roy Smith, Hank Whittemore, Ursula Goodenough, Patrick Huyghe, Brian Hayes, Mary Flynn, Jim DeVinney, Tim Hays, Lisa Fuerst, Bill O'Reilly, Gina Webster, Jim Mairs, Eleanor Kostyk, Elliott Erwitt, Tara Kuellhoffer, Diana Lutz, Karen Auerbach, Nancy Lattanzio, Nancy Eisenbarth, Todd Brewer, Noel Buckner, Rob Whittlesey, Karen Johnson, Rob Harte, and Emily Loose. And not least, for a man who is very much a stranger to his own stove and oven, I'm indebted to Jimmy Ray, Pat Robustelli, and Anthony Perroncino of JR's Restaurant, the DiScala family of Peppino's, Thomas Sprague of Bob's Diner, and the O'Leary family of The Blazer.

Photo Credits

Chapter 1:	Photo courtesy of the California Institute of Technology.
Chapter 2:	Photo courtesy of Einstein Archives/AIP Emilio Segrè Visual Archives.
Chapter 3:	Photo used by permission of the Nobel Foundation.
Chapter 4:	Photo courtesy of Mary Evans Picture Library.
Chapter 5:	Photo courtesy of AIP Emilio Segrè Visual Archives.
Chapter 6:	Photo courtesy of the Institute for Advance Studies Archives.
Chapter 7:	Photo courtesy of Photo Researchers.
Chapter 8:	Photo used by permission of Brown Brothers.
Chapter 9:	Photo courtesy of the Massachusetts Institute of Technology Museum.
Chapter 10:	Photo courtesy of Corbis Corporation.

Index

INDEX